D0778789

MINDS MADE FEEBLE:

Deborah at the Sewing Machine. Reprinted from *The Kallikak Family: A Study in the Heredity of Feeble-Mindedness* by H.H. Goddard, 1912, New York: Macmillan.

MINDS MADE FEEBLE:

The Myth and Legacy of the Kallikaks

J. David Smith

AN ASPEN PUBLICATION®

Aspen Systems Corporation
Rockville, Maryland
Royal Tunbridge Wells
1985

Library of Congress Cataloging in Publication Data

Smith, John David, 1949 –
Minds made feeble.

"An Aspen publication."
Bibliography: p.
Includes index.
1. Eugenics. 2. Mental retardation. 3. Heredity, Human.
4. Goddard, Henry Herbert, 1866 – 1957. The Kallikak family.
I. Title. II. Title: Kallikaks.
HQ755.35.S65   1985   304.5   84-28416
ISBN: 0-87189-093-3

Managing Editor: M. Eileen Higgins
Printing and Manufacturing: Debbie Collins

Library of Congress Catalog Card Number: 84-28416
ISBN: 0-87189-093-3

Printed in the United States of America

1   2   3   4   5

To
Joyce

and to our children,
Lincoln, Allison, and Sallie.

Contents

Foreword   *ix*
Acknowledgments   *xiii*

Chapter 1—Introduction   1
Chapter 2—The Study   11
Chapter 3—Deborah   21
Chapter 4—The Context   37
Chapter 5—Elizabeth Kite   49
Chapter 6—Acclaim, Criticism, and Defense   61
Chapter 7—Revisiting the Kallikaks   83
Chapter 8—Immigrants, Morons, and Democracy   115
Chapter 9—Eugenics, Sterilization, and the Final
            Solution   135
Chapter 10—The New Eugenics   169

Epilogue   *191*
Index   *195*

Foreword

IT IS STRANGE THAT SO MANY YEARS WENT BY BEFORE SOME-
one tried to find out the real name of the Kallikak family,
making it possible to test Goddard's descriptions and conclusions against reality. After all, Goddard's work not only
gave rise to considerable scientific controversy, but also had
a profound influence in matters of public policy, legislation,
institution building, and public education. With each passing decade the critiques of Goddard's study multiplied, and
even those who were "hereditarians" agreed that his study
was a frail reed to which to tie their arguments. However, it
seemed as if we would never know what the ignoble

Kallikaks were really like—where and how they lived; the ways they conducted their vocational, personal, and familial lives; and how their contemporaries saw them.

What Dr. Smith presents in this book is a fascinating example of "psychological archeology." Archeology involves digging—literal or symbolical—and taxes one's capacity to endure tedium. Motivation comes from the knowledge that if one hits pay dirt the intellectual rewards are beyond measure. Dr. Smith has hit pay dirt, and science in general, and psychology in particular, are in his debt. Before he wrote this book we knew that Goddard had vastly overstated his case, but now we know it with a degree of certainty that is both heartening and surprising.

This is a book for everyone: for those who need to be reminded that science is a human enterprise, for those who oversimplify the relationship between data and interpretation, and those who want their horizons about our society's past and present enlarged, for those who want better to understand how prejudice flows through national and international networks, and for those who need to be shocked into recognizing how the issues surrounding the Kallikak family are still with us today.

I first came to know about the Kallikaks in 1935 in a psychology course at the University of Newark (now the Rutgers campus there). I learned a lot about them because my professor, Dr. Gaudet, disagreed completely with Goddard's methods and conclusions. Also, not unimportantly, the Kallikaks were a New Jersey family. I learned even more about them, from a somewhat contrary perspective, in a special course on mental deficiency taught by Dr. Lloyd Yepsen, then the Chief Psychologist in New Jersey's State Department of Institutions and Agencies. As the years went on I read whatever I could find on the Kallikaks, and in several of my books, alone and together with Dr. John Doris (another psychological archeologist), critiqued Goddard's publications.

It is not hyperbole to say that a library of modest size would be required to house all that has been written about that family. Interest in them, far from declining, seems to have remained constant or to have increased—much like the situation in regard to Jean-Marc-Gastard Itard's early 19th century studies of the *Wild Boy of Aveyron.* As a result of these studies, our knowledge and perspective have changed, but not in truly dramatic ways. However, after reading Dr. Smith's book, I found both my knowledge and perspective discernibly altered, not in the sense of changing my basic position but in the sense of clarity about two things: first, how utterly wrong Goddard was, and second, how eager some individuals and groups are to use pseudo-science as a basis for political agendas that deny opportunity—and even life itself—to certain segments of society. Just as today (see Joan Peter's *From Time Immemorial*) there are people in power on this earth who parade as fact the forged "Protocols of the Elders of Zion," so there are people who continue to seize on the Goddard study as scientific truth. If the minds of such people cannot be influenced, let us at least not underestimate their presence, numbers, and, not infrequently, their positions of power.

Dr. Smith's book is both illumination and warning, testimony to what is best and worst in social living and a reminder that between "data" and social action lies a mine field that few have traversed without causing harm to themselves or others. The saga of the Kallikaks continues to unfold. There is and should be no final chapter.

SEYMOUR B. SARASON
Professor of Psychology
Yale University
New Haven, Connecticut

Acknowledgments

A S A CHILD I OFTEN WONDERED WHY GRACE WAS SAID BEFORE meals rather than after. It seemed more reasonable to me that thanks should be withheld until it was certain that it was deserved by the quality of the meal. A number of years passed before I understood that giving thanks for what is yet to be received is a matter of faith and trust. It is appropriate that my opportunity to say thanks to the many people who helped with this book comes at the beginning. These are people who had the grace to help me with this effort long before they could know if their efforts would be justified. They acted on the basis of trust, and I will always appreciate their faith in the importance of what I was attempting.

The research for this book brought me into contact with several fine institutions and many knowledgeable and generous people. A special note of thanks is due to the staff of the Archives of the History of American Psychology and to the founders and life forces of that fine facility: *John Popplestone* and *Marion White McPherson.* Staff members at the Library of Congress and the National Archives provided invaluable assistance. Through using these two resources during my research on the Kallikaks, other new worlds of information were opened to me. How fortunate we are to have these wonderful repositories.

From the beginning of my work on this book I had the encouragement and support of my colleagues, *Ed Polloway* and *Ken West.* Even when my research took unusual and obscure turns, they kindly accepted my obsessions. *Rosel Schewel* has been a quiet source of reassurance to me for almost a decade. Her interest in my ideas and my work is greatly valued. *Pete Warren* has been a constant friend and an unfailing wellspring of care. *Deborah Beckel,* archivist and research librarian at the Lynchburg College Library, believed my idea would become a book long before I was really convinced. Her knowledge and skill were integral to locating many of the sources cited in this book. For that assistance and for her editorial comments on the early drafts of chapters I am thankful. Thanks also go to *Carol Pollock* of the library for her reference help. Lynchburg College, through the Committee on Faculty Research and Development, provided some financial support for my research. That help was welcomed and I continue to appreciate it as an expression of confidence in my work.

I was received with courtesy and kindness in many courthouses, libraries, and historical societies in New Jersey. Each bit of information I found in these places added to my comprehension of the Kallikak family. My gratitude goes out to those who helped me find some of the smaller, but critical, pieces of the puzzle. A special thank you is due *Roxanne Carkhuff.* Her expertise shortened my search by several light years. Her skill in genealogical research was invaluable and her enthusiasm for the project was encouraging.

As I was writing this book I asked several people to read portions of the manuscript. These are all people whom I respect and trust. Thanks for helpful criticism and enlightening comments to *Len Blackman, Bob Bogdan, Ignacy Goldberg, Stephen Jay Gould, Bob Lassiter, Maggie Noel,* and my brother, *Carl Smith.*

Major credit for this book reaching publication belongs to *Burton Blatt* and *Seymour Sarason.* Both of these men spoke on behalf of the book in words that were humbling to me and generated interest in others. I shall always remember the special consideration they took with the manuscript and the unsolicited words of endorsement and advice each offered. I hope that in the future I shall have the capacity to emulate the selfless advocacy and scholarly encouragement given to me by Professor Blatt and Professor Sarason.

Betty Johnson typed the manuscript with care. She, however, did much more than typing. Through her vigilance, many errors of form and expression were corrected. Her cheerful acceptance of my handwriting is a testament of tolerance.

Finally, I am grateful that *Margaret Quinlin* and her associates at Aspen Systems decided to take on a book that is "special." The concern that has been shown in achieving the best quality possible in this book has won my admiration.

Chapter 1

Introduction

EDUCATION "HAS A KIND OF INTELLECTUAL AND MORAL OMNI-potence; that to its different forms are to be ascribed the chief, if not all the differences observable in the genius, talents, and dispositions of men; and that by improving its principles and plan, human nature may, and finally will, reach a state of absolute perfection in this world, or at least go on to a state of unlimited improvement." This assessment of the American faith in education was written by Samuel Miller, a British subject, in his study of the developing culture of the United States following the Revolution. He went on to write that Americans held that "man is the child of circumstances; and by meliorating these . . . his true and highest elevation is to be obtained; and they even go so far as to believe that, by means of the advancement of light and knowledge, all vice, misery, and death may finally be banished from the earth" (cited in Cremin, 1970, p. 562).

Although Miller's critique of the new nation's confidence in education is somewhat satiric, it does describe a belief that has been woven into the fabric of American social thought. This belief—the malleability of the human mind, personality, and condition—has prevailed to some degree throughout our history as a nation. Indeed, the idea that people are largely the products of experience, education, and opportunity has had a continuing impact upon the nature of our politics, social institutions, and educational system.

Our history, however, has also been influenced by the contrary concept that portions of the human race are inherently and genetically inferior, and that this status is not modifiable. One manifestation of this point of view was social Darwinism, which arose as a philosophic and scientific movement during the late nineteenth century. The movement hinged on the idea that certain racial and ethnic groups are inherently inferior in intelligence and moral character and that, even within cultural or national groups, lower social classes are, by nature, inferior. Social Darwinism was used as a justification for colonization by the dominant imperial powers, which were simply assuming the "white man's burden" of responsibility for controlling, protecting, and bringing salvation to inferior races. Industrialists combated attempts at unionization of their workers by arguing that the "common man" was incapable of determining what was best for himself.

In 1883, Francis Galton, cousin of Charles Darwin, introduced the term *eugenics,* which he defined as the science that would deal with all of the influences that could improve the inborn qualities of a race. An early eugenic aim was the elimination from human populations of unwanted hereditary disorders by selective marriage practices. Quickly, however, the movement spread to encompass not only the promotion of compulsory sterilization of people with undesirable traits but also the restricted immigration of unwanted races and nationalities—groups that by eugenic definition possessed inferior hereditary material.

The eugenicists had powerful political allies who helped in advancing their cause. In the United States, although some scientists spoke out against the questionable research being done in the name of human improvement, the eugenicists successfully lobbied for compulsory sterilization laws in some 30 states. The Immigration Restriction Act of 1924 (which remained in effect until 1965) was passed largely because of supporting testimony provided by the staff of the Eugenics Record Office of Cold Spring Harbor on Long Island, the center of power and influence of the American eugenics effort.

The eugenicists concentrated much of their research effort on human pedigrees. They argued that most mental retardation was genetic and could be found occurring generation after generation in certain families. The results of their studies provoked emotionally charged pronouncements from the eugenicists. In his book, *The Revolt Against Civilization,* Lothrop Stoddard (1922) stated that the uncontrolled reproduction among defective families and the intermingling of defective and normal human stock was resulting in the "twilight of the American mind," the "dusk of mankind." He argued that "in former times the numbers of the feebleminded were kept down by the stern processes of natural selection, but modern society and philanthropy have protected them and thus favored their rapid multiplication" (p. 94).

Although there were many studies of family degeneracy (the Jukes and Nams of New York, the Tribe of Ishmael in Indiana, the Hill Folk of Ohio, and the Dacks of Pennsylvania), the most powerful and influential was reported in 1912 by Henry Goddard in his book *The Kallikak Family: A Study in the Heredity of Feeble-Mindedness.* It is that study—its genesis and its legacy—that is the reason for and the concern of this book.

Goddard based his study on the family background of a young girl to whom he gave the pseudonym Deborah Kallikak. She was a resident at the Training School for Feeble-Minded Girls and Boys in Vineland, New Jersey. A description of the study and what Goddard thought he had found in Deborah, her relatives, and her ancestors is the topic of chapter 2 of the present volume. He was convinced that his findings proved that mental retardation was almost always a matter of tainted blood—of a bad seed. Moreover, he interpreted his data on the Kallikak family as evidence that prostitution, alcoholism, criminality, and other social ills were merely byproducts of the same genetic flaw that caused retardation. Since Deborah was the starting point for his research, it was essential that Goddard's diagnosis of her feeblemindedness was correct. Chapter 3 describes Deborah and offers glimpses of her long life of institutionalization. The reader is invited to question the accuracy of Deborah's classification as feebleminded and the necessity for her life of confinement.

Chapter 4 is an exploration of the factors that led to Goddard's search for a genetic explanation of mental retardation. The conservative social climate of the time and the emerging popularity of eugenic principles in scientific circles provided a hospitable atmosphere for the direction in which he took his work. More important, however, seemed to be the personal influence of the teachers, colleagues, and professional acquaintances who advised and supported him in the work.

The information on the Kallikak family was actually collected by one of Goddard's field workers, Elizabeth Kite. She was assigned the task of finding court records and medical histories of Deborah's relatives and ancestors. Incredibly, Goddard also relied upon her to make diagnostic decisions about the mental ability and personality characteristics of members of the family, both living and dead. Elizabeth Kite was not a psychologist or physician; nor was she trained in sociology, anthropology, or any other social science. But Goddard believed that, after a few weeks of experience and instruction at the training school, his field workers became proficient at recognizing the presence and degree of feeble-mindedness in those they were dispatched to study. In relation to those people who were dead or unavailable but on whom his field workers had to arrive at a diagnosis, he said:

> Some record or memory is generally obtainable of how the person lived, how he conducted himself, whether he was capable of making a living, how he brought up his children, what his reputation was in the community; these facts are frequently sufficient to enable one to determine, with a high degree of accuracy, whether the individual was normal or otherwise. (Goddard, 1912, p. 14)

Chapter 5 is an examination of Elizabeth Kite's involvement in the study and the methods she used in tracing Deborah's heritage.

Goddard's book on the Kallikak family was received with acclaim by the public and by much of the scientific community. It went through several editions. Overtures were made to Goddard concerning the possibility of a Broadway production based on the book. It was given very favorable reviews in both popular magazines and scientific periodicals. Only gradually did criticism arise concerning the methods used in the study and the implications and conclusions drawn from the data collected. Even in the light of substantive and knowledgeable criticism, however, the essential message of the Kallikak study persisted for years. Even today, in convoluted forms, it continues to have a social and political impact. Its message is simple, yet powerful. Ignorance, poverty, and social pathology are in the blood—in the seed. It is not the environment in which people are born and develop that makes the critical difference in human lives. People are born either favored or beyond help. Social programs, "wars on poverty," and compensatory education are futile and wasteful. Chapter 6 examines the popularity of the Kallikak study, its critics, and its defense by Goddard and his friends.

I have known of the Kallikak study since I was in college. While aware even then of its methodological weaknesses and the questionable assumptions it made from the data, I have continued to wonder about the lives of the real people described in the book. Later, while teaching my own students to dismiss the study as bad research with biased conclusions, I was still troubled by images of Deborah and her relatives. Even though an environmental explanation of the family's generations of poverty, retardation, and social ills could easily be substituted for Goddard's hereditary explanation, had all the "bad" Kallikaks actually been as degenerate as they had been portrayed? In chapter 7 I describe my discovery of the "real" Kallikaks. The historical evidence shows them in fact to be victims of a philosophic and pseudoscientific movement. The truth of their lives was sacrificed to an effort to prove a point. The Kallikak study is fiction draped in the social science of its time. This is the most important chapter of the book.

Following his investigation of Deborah Kallikak's geneal-ogy and similar studies of other residents at the Vineland training school, Goddard turned his attention to the question of immigration. Under his direction, newly arrived immi-grants were tested to determine their level of intelligence. Based on the results of this work on Ellis Island, he came to the conclusion that most of the immigrants entering the United States were of low intelligence. He estimated that the average immigrant had an intelligence level of "the moron grade" (Goddard, 1917, p. 243). He totally rejected the idea that his tests might be biased against foreign newcomers to this country or that there might be other physical or psycho-logical factors—for example, fear or fatigue—that influ-enced their results. Reports of Goddard's research contributed to the increased deportation of immigrants for reason of mental deficiency and the passage in 1924 of the Immigration Restriction Act. These events, along with a discussion of Goddard's views on democracy, are the subject of chapter 8.

The Kallikak study, complemented by Goddard's subsequent work and that of other eugenicists, proved to be a very potent indictment of the poor, the uneducated, racial minorities, the foreign-born, and those classified as mentally retarded or mentally ill. The study was used by the privileged to justify the naturalness of their privileges—only the "good stock" was capable of acquiring and managing power and prerogatives. In addition to enacting compulsory sterilization laws, restricting immigration, and creating more and larger institutions for those persons deemed deficient or defective, politicians could argue on this basis against the expenditure of funds for education, health, and housing for the "Kallikaks" of the land. According to Goddard (1912),

> they were feeble-minded and no amount of education or good environment can change a feeble-minded individual into a normal one, anymore than it can change a red-haired stock into a black-haired stock. The striking fact of the enormous proportion of feeble-minded individuals in the descendants of Martin Kallikak, Jr. and the total absence of such in the descendants of his half brothers and sisters is conclusive on this point. Clearly it was not environment that has made that good family. They made their environment; and their own good blood, with the good blood in the families into which they married, told. (p. 53)

Ultimately, the eugenic concepts that the Kallikaks were used to illustrate evolved into the racial hygiene program of Nazi Germany. From the Kallikaks to the Holocaust may seem a fantastic leap, but the connection between the American eugenics movement and Hitler's "final solution" is clear and will be discussed in chapter 9.

For many years, the Kallikak story was presented and accepted as proof of the inheritance of good and bad human traits. Generations of college and graduate students were influenced by it in their personal and professional lives. Even after extensive criticism of the study had been published, it was often discussed as research that, although methodologically flawed, was benign, as research that simply left open the question of nature versus nurture in determining such human characteristics as intelligence. Indeed, it was at times defended as a pioneering effort which, although technically primitive, was sound in the questions it posed and valid in its general findings.

The horrors exposed at the Nuremberg Trials dampened the fervor of the American eugenics movement and, with it, the prominence of the Kallikak study. However, the themes expressed in the story have been reincarnated in recent years—sometimes in a more subtle and sophisticated form and, remarkably, sometimes in almost identical fashion. Awareness of the link between the "new eugenics" and the old is important. This is the purpose of chapter 10.

While working on this book, I have been fortunate in having friends and colleagues encourage me with assurances of the significance of the effort. Occasionally, however, in explaining the project to others, I have detected a glaze of incredulity in their expressions of interest. A couple of times I have actually been asked why I would want to devote time and effort to a 75-year-old study that is now widely acknowledged as invalid. In moments of doubt and discouragement I have asked myself the same question.

Yet, I am convinced, more now than when I started, that the complete story of the Kallikak family needed to be found and told. It is, perhaps more than anything else, an illustration of the power of a social myth. Goddard found the characters that he could make fit the tale; the Kallikaks gave human form to a story that the social Darwinists and eugenicists had been developing for decades. The public found in the book a parable they wanted to believe. Politicians, policy makers, and the otherwise powerful found evidence in the Kallikak family that their disregard of the rights of the weak was consistent with the natural order of life and in the best interest of the nation.

Social myths are constantly in the making—compelling in their simplicity, and alluring because we want to believe them. Perhaps understanding the Kallikak story will help in recognizing and resisting such myths. I believe this will be true for me.

Finally, I hope that this book will reveal for some, and remind others of, the tragedy of Deborah Kallikak's life of needless confinement, and of the thousands of other lives that have been similarly wasted. If the pages that follow accomplish that to some degree, the effort in producing them will have been justified.

REFERENCES

Cremin, L. (1970). *American education: The colonial experience 1607–1783*. New York: Harper & Row.

Goddard, H.H. (1912). *The Kallikak family: A study in the heredity of feeble-mindedness*. New York: Macmillan.

Goddard, H.H. (1917). Mental tests and the immigrant. *Journal of Delinquency, 2*, 243–277.

Stoddard, L. (1922). *The revolt against civilization*. New York: Charles Scribner's Sons.

Chapter 2

The Study

I<small>N</small> 1912, H<small>ENRY</small> H<small>ERBERT</small> G<small>ODDARD</small>, <small>DIRECTOR OF THE</small> research laboratory of the Training School for Feeble-Minded Girls and Boys in Vineland, New Jersey, published his account of a family that had come to his attention in the course of investigating the role of heredity in mental retardation. The study of the family tree had begun with a young woman who was a resident of the Vineland institution. Deborah Kallikak was considered to be "feebleminded." More specifically, she had been classified as a *moron,* a designation that Goddard had coined from a Greek word meaning foolish. The label moron came to be widely applied to peo-

ple who were considered to be "high grade defectives"—
those who were not retarded seriously enough to be obvious
to the casual observer and who had not been brain-
damaged by disease or injury. Morons were characterized
as being intellectually dull, socially inadequate, and mor-
ally deficient. From the beginning of his research, Goddard
was inclined to believe that these traits were hereditary in
origin. He was of the opinion that reproduction among people with these traits posed a threat to the social order and the
advancement of civilization.

Deborah, who had been born in an almshouse, was
admitted to the Vineland training school at the age of 8. She
was almost 23 when the study was published. The name
Deborah and the family name Kallikak which she, her relatives, and her ancestors were given are pseudonyms. Goddard seems to have enjoyed inventing terms; he composed
Kallikak from the Greek words *kallos* (beauty) and *kakos*
(bad). He used this composite as a symbol of the two hereditary influences that he believed had resulted in Deborah's
moronity. A good, or beautiful, hereditary strain in her background had been tainted by a bad genetic seed, ultimately
producing her inferior intellect. Although the names of the
family members were fictitious, Goddard emphasized that
the "present study of the Kallikak family is a genuine story of
real people" (Goddard, 1912, p. viii).

When Goddard was hired by the Vineland training school
in 1906, his primary charge was to conduct research that
might lead to the discovery of the causes of feeblemindedness. This challenge eventually resulted in his focus on those
"high grades," the morons, where a medical explanation for
retardation was not apparent. It also led to a research strategy:

After some preliminary work, it was concluded that the
only way to get the information needed was by sending
trained workers to the homes of the children, to learn by
careful and wise questioning the facts that could be

obtained. It was a great surprise to us to discover so much mental defect in the families of so many of these children. The results of the study of more than 300 families will soon be published, showing that about 65 per cent of these children have the hereditary taint. (Goddard, 1912, p. viii)

In 1914, Goddard presented the results of the complete project in the book, *Feeble-Mindedness: Its Causes and Consequences.* He included selected case studies and a summary of the investigation of the 300 families. A special aspect of the study of the Kallikak family, however, pushed Goddard to the earlier publication, in 1912, of the book, *The Kallikak Family: A Study in the Heredity of Feeble-Mindedness.*

A field worker named Elizabeth Kite was assigned by Goddard to investigate Deborah's family. Through interviews and observations of her living relatives, Kite discovered that,

after Deborah's birth in the almshouse, the mother had been taken with her child into a good family. Even in this guarded position, she was sought out by a feeble-minded man of low habits. Every possible means was employed to separate the pair, but without effect. Her mistress then insisted that they marry, and herself attended to all the details. After Deborah's mother had borne this man two children, the pair went to live on the farm of an unmarried man possessing some property, but little intelligence. The husband was an imbecile who had never provided for his wife. She was still pretty, almost girlish—the farmer was good-looking, and soon the two were openly living together and the husband had left. As the facts became known, there was considerable protest in the neighborhood, but no active steps were taken until two or three children had been born. Finally, a number of leading citizens,

Illustration 1: Deborah Kallikak circa 1912. Reprinted from *The Kallikak Family: A Study in the Heredity of Feeble-Mindedness* by H.H. Goddard, 1912, New York: Macmillan.

headed by the good woman before alluded to, took the matter up in earnest. They found the husband and persuaded him to allow them to get him a divorce. Then they compelled the farmer to marry the woman. He agreed, on condition that the children which were not his should be sent away. It was at this juncture that Deborah was brought to the Training School. (Goddard, 1912, pp. 64–65)

This description is typical of those of family backgrounds found for many of the children at the training school. While such information could be used to support the hereditary view that mental inferiority reproduces itself, an equally potent argument could be made that deficient and disturbed environments create people who are intellectually and socially defective.

Illustration 1 is a photograph of Deborah. It was taken at about the time of the study.

The distinguishing features of the Kallikak family were garnered from historical documents and the recollections of elderly informants. Through the mother, Deborah's genealogy was eventually traced back to her great-great-grandfather, Martin Kallikak. Elizabeth Kite reported that in each generation the family was characterized by deficiency and degeneracy:

The surprise and horror of it all was that no matter where we traced them, whether in the prosperous rural district, in the city slums to which some had drifted, or in the more remote mountain regions, or whether it was a question of the second or the sixth generation, an appalling amount of defectiveness was everywhere found. (Goddard, 1912, p. 16)

In the course of the investigation, Miss Kite repeatedly came across another Kallikak family that apparently was not related to Deborah or her ancestors. This family was from an upstanding and prosperous line; it was a family esteemed in its community and noted as being of good stock. Gradually Goddard became convinced that Deborah's genealogy was a degenerate offshoot of the better line.

Then, through Miss Kite's persistent sleuthing, came an unexpected discovery that would make the Kallikak family unique in the hereditary research done at Vineland:

> The great-great-grandfather of Deborah was Martin Kallikak. That we knew. We had also traced the good family, before alluded to, back to an ancestor belonging to an older generation than this Martin Kallikak, but bearing the same name. He was the father of a large family. His eldest son was named Frederick, but there was no son by the name of Martin. Consequently, no connection could be made. Many months later, a granddaughter of Martin revealed, in a burst of confidence, the situation. She told us (and this was afterwards fully verified) that Martin had a half brother Frederick—and that Martin never had an own brother "because" as she now naively expressed it, "you see his mother had him before she was married." (Goddard, 1912, p. 17)

From this revelation flows the story that has proved to be an intriguing and powerful social myth in western cultures during much of the twentieth century. Upon reaching young manhood, the senior Martin Kallikak joined the militia; this was during the early days of the American Revolution. At a New Jersey tavern frequented by soldiers, he met a woman whom almost a century and a half later Goddard diagnosed as feebleminded. From their relationship a son was born; the son was also reported to be feebleminded. The mother gave the child Martin's full name. According to Goddard,

> this illegitimate boy was Martin Kallikak, Jr., the great-great-grandfather of our Deborah, and from him have come four hundred and eighty descendants. One hun-

dred and forty-three of these, we have conclusive proof, were or are feeble-minded, while only forty-six have been found normal. The rest are unknown or doubtful. (Goddard, 1912, p. 18)

Upon leaving the militia, Martin, Sr., returned home and married a respectable woman from a good family. From this union the upstanding line described by Elizabeth Kite descended:

All of the legitimate children of Martin Sr. married into the best families in their state, the descendants of colonial governors, signers of the Declaration of Independence, soldiers and even founders of a great university. Indeed, in this family and its collateral branches, we find nothing but good representative citizenship. There are doctors, lawyers, judges, educators, traders, landholders, in short, respectable citizens, men and women prominent in every phase of social life. They have scattered over the United States and are prominent in their communities wherever they have gone. Half a dozen towns in New Jersey are named from the families into which Martin's descendants have married. (Goddard, 1912, pp. 29–30)

Thus we are presented with a seemingly perfect dichotomy. Martin, Sr.'s, liaison with the tavern's "nameless girl" resulted in continuing generations of defective people; his marriage to a woman of good stock produced human excellence in an undisturbed progression. Goddard felt that the discovery of the Kallikaks "presented a natural experiment in heredity." Here was evidence that, because of the difference in the quality of the two women with whom Martin, Sr., had fathered children, one line was thoroughly good and the other was riddled with a defective trait:

This defect was transmitted through the father in the first generation. In later generations, more defect was brought in from other families through marriage. In the last generation it was transmitted through the mother, so that we have here all combinations of transmission, which again proves the truly hereditary character of the defect.

We find on the good side of the family prominent people in all walks of life and nearly all of the 496 descendants owners of land or proprietors. On the bad side we find paupers, criminals, prostitutes, drunkards, and examples of all forms of social pest with which modern society is burdened. (Goddard, 1912, p. 116)

Goddard concluded that feeblemindedness was at the core of the cited and other social problems. More precisely, he identified the moron as the source of these problems and the target for their solution. It is the moron who reproduces more and more of his type to become a drain and a danger in society:

We have the type of family which the social worker meets continually and which makes most of our social problems. A study of it will help to account for the conviction we have that no amount of work in the slums or removing the slums from our cities will ever be successful until we take care of those who make the slums what they are. Unless the two lines of work go on together, either one is bound to be futile in itself. If all of the slum districts of our cities were removed tomorrow and model tenements built in their places, we would still have slums in a week's time because we have these mentally defective people who can never be taught to live otherwise than as they have been living. Not until we take care of this class and see to it that their lives are guided by intelligent people, shall we remove these sores from our social life.

There are Kallikak families all about us. They are multiplying at twice the rate of the general population, and not until we recognize this fact, and work on this basis, will we begin to solve these social problems. (Goddard, 1912, pp. 70–71)

"What is to be done?" Goddard asks rhetorically. He answers by acknowledging that compulsory sterilization might be used as a temporary and emergency measure. He sees segregation of the feebleminded by institutionalization, however, as the best solution:

If such colonies were provided in sufficient number to take care of all the distinctly feeble-minded cases in the community, they would very largely take the place of our present almshouses and prisons, and they would greatly decrease the number in our insane hospitals. Such colonies would save an annual loss in property and life, due to the action of these irresponsible people, sufficient to nearly, or quite, offset the expense of the new plant. . . . Segregation through colonization seems in the present state of our knowledge to be the ideal and perfectly satisfactory method. (Goddard, 1912, pp. 105–106, 117)

REFERENCES

Goddard, H.H. (1912). *The Kallikak family: A study in the heredity of feeble-mindedness.* New York: Macmillan.

Goddard, H.H. (1914). *Feeble-mindedness: Its causes and consequences.* New York: Macmillan.

Chapter 3

Deborah

Deborah's presence at the Vineland training school was fundamental to Goddard's study of the Kallikak family—fundamental in the sense, of course, that she was the starting point for his odyssey back through the Kallikak generations. More importantly, Deborah's residence at Vineland and the diagnosis of her as feebleminded was basic to the argument of the "bad seed." To Goddard she served as the central example of the continuing and inevitable influence of hereditary mental defects. Accordingly, he describes Deborah in detail in order to confirm her status as a moron.

A review of these descriptions today, however, casts considerable doubt on Goddard's diagnosis and the necessity for Deborah's institutionalization. The following excerpts are from the admission information of November, 1897, when Deborah was 8 years old:

> Average size and weight. No peculiarity in form or size of head . . . washes and dresses herself, except for fastening clothes . . . knows all the colors. Not fond of music . . . can use a needle . . . careless in dress . . . obstinate and destructive . . . does not mind slapping and scolding. (Goddard, 1912, p. 2)

The reports indicate that, at age ten, Deborah could do some reading, writing, and counting but that her conduct was "quite bad—impudent and growing worse." By 1900, when she was 11, the reports contradict the earlier description of her lack of fondness for music:

> Good in entertainment work. Memorizes quickly. Can always be relied upon for either speaking or singing. Marches well . . . knows different notes. Plays "Jesus, Lover of my Soul" nicely. Plays scale of C and F on cornet. (Goddard, 1912, p. 3)

The reports of 1901 include the following comments:

> She plays by ear. She has not learned to read the notes . . . simply because she will not put her mind to it. She has played hymns in simple time, but the fingering has had to be written for her . . . excellent worker in gardening class. Has just completed a very good diagram of our garden to show at Annual Meeting . . . has nearly finished outlining a pillow sham . . . is very good in number work, especially in addition . . . is restless in class. Likes to be first in everything . . . she could learn more in school if she would pay attention, but her mind

seems away off from the subject in discussion. (Goddard, 1912, pp. 3–4)

By age 15 (see Illustration 2), Deborah had become quite skillful with a sewing machine and was making some of her own clothing. She continued to play the cornet and had learned to read music. Her conduct was evaluated as "fair."

Four years later, the reports listed defects primarily in academic areas, indicated her conduct had improved, and continued to praise her talent in crafts and artistic works:

Drawing, painting, coloring, and any kind of hand work she does quite nicely . . . this year she had made a carved book rest with mission ends and is now working on a shirtwaist box with mortise and tenon joints and lap joints. The top will be paneled. (Goddard, 1921, p. 5)

At age 20, the following was recorded:

Made the suit which she had embroidered earlier in the year, using the machine in making it. Helped F.B. put her chair together and really acted as a teacher in showing her how to upholster it. Will be a helper in wood-carving class this summer. (Goddard, 1912, p. 6)

In 1911, the year before the Kallikak book was published, 22-year-old Deborah was described as a skillful and hard worker who lacked confidence in herself. She continued to excel in woodworking and dressmaking. Academic subjects were still a problem: "Can write a well-worded story, but has to have more than half the words spelled for her" (Goddard, 1912, p. 6). Samples of Deborah's handiwork at this time are shown in Illustration 3.

When first reading the accounts, from which these excerpts were taken, of Deborah's progress at the training school, I was struck with the similarity between them and numerous diagnostic profiles of learning disabled children

Illustration 2: Deborah Kallikak at age 15. Reprinted from *The Kallikak Family: A Study in the Heredity of Feeble-Mindedness* by H.H. Goddard, 1912, New York: Macmillan.

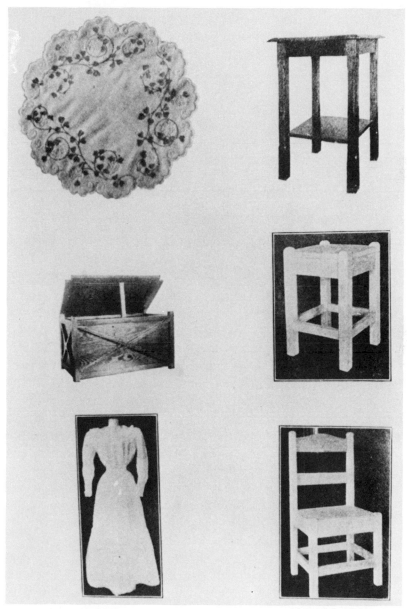

Illustration 3: Samples of Deborah's handiwork at the Vineland Training School. Reprinted from *The Kallikak Family: A Study in the Heredity of Feeble-Mindedness* by H.H. Goddard, 1912, New York: Macmillan.

and young adults that I have encountered. The classic picture of language-related difficulties and marked strength in nonverbal areas emerges. It is very likely that many psychologists and educational diagnosticians today would view Deborah's difficulties from a learning disabilities perspective rather than as the result of being mentally retarded.

Also, given Deborah's troubled early life, every consideration would have to be given to possible emotional components in her learning problems and social difficulties. Following her birth in an almshouse, she had lived with her mother in circumstances that, from Goddard's description, seemed to have been characterized by social, economic, and physical flux. Her mother had married on the condition that she send away the children who were not the offspring of her new husband, thus sending Deborah to the training school. The early reports and later glimpses of her life characterize Deborah as an easily disturbed and emotionally volatile person. The degree to which her difficulties were emotional rather than intellectual is certainly open to question.

The fact that Deborah was born into a tradition of economic deprivation and social isolation must also be taken into consideration. Goddard's descriptions of her ancestry, if viewed from an environmental perspective, portray a family that for generations had existed in poverty. There were apparently few opportunities for formal education for her predecessors. There are indications that the family had always lived in isolation from the surrounding culture. At the training school, Deborah was evaluated according to standards and values that were alien to what her early life experience had taught her.

Still, the records indicate that Deborah was learning and growing during those years at the training school. Each year seemed to bring development in her life, particularly in nonacademic learning and in social skills. It might in fact be argued that her institutionalization was justified: She was learning, becoming a more functional person, and achieving personal independence. Perhaps institutionalization would equip her for a later productive life in society.

Deborah, however, would never be a member of a society other than that of an institution. She was destined to live a total of 81 years in two institutions. From the time she entered the training school until she died at the Vineland State School across the street, she would never know life free of institutional influence. When she died in 1978, she was buried in the institution's cemetery under a marker bearing only her name.

Descriptions of Deborah subsequent to the publication of the Kallikak study repeatedly refer to her beauty and charm. A photograph of her at age 17 is shown in Illustration 4. In 1983, Eugene Doll, son of Edgar Doll who worked with Goddard as an assistant from 1912 to 1917 and who in 1925 became director of research at Vineland, wrote:

> There is no doubt that, whatever her mentality, she radiated that extra spark of personality which makes one stand out in a crowd and which not only attracts but holds friends. J.E. Wallace Wallin wrote urbanely of his first encounter with Deborah—finding her in charge of the kindergarten at the Training School and mistaking her for the teacher. At lunchtime he was surprised to find the same attractive young woman waiting on his table. . . .[1]
>
> Time and again visitors in both the Training School and the Vineland State School . . . to which Deborah was later transferred, commented on her seeming normality. (Doll, 1983, p. 30)

Helen Reeves, executive social worker at the Vineland State School, commented on Deborah's transfer from the training school:

> For our part we knew we had acquired distinction in acquiring Deborah Kallikak, for by this time the story of

Illustration 4: Deborah Kallikak at age 17. Reprinted from *The Kallikak Family: A Study in the Heredity of Feeble-Mindedness* by H.H. Goddard, 1912, New York: Macmillan.

her pedigree was becoming well known. And such a capable, well trained and good looking girl must be an asset. . . . Deborah at this time was a handsome young woman, twenty-five years old, with many accomplishments, though her academic progress had remained stationary just beyond second grade. She excelled in the manual arts of embroidery, woodcraft and basketry, played the cornet beautifully and took star roles in all entertainments as a matter of course. She was well trained in fine laundry work and dining room service, could use a power sewing machine and had given valuable assistance as a helper in cottages for low grade children. Her manner toward her superior officers was one of dignified courtesy. (Reeves, 1938, pp. 195–196)

Deborah was given special responsibilities during most of her life. As an adolescent, she served in the home of the superintendent of the training school. In addition to performing housekeeping duties, she cared for the family's infant son. She later assumed child care responsibilities for the assistant superintendent of the state school. Children from both of these families continued to visit and correspond with Deborah throughout her life. A woman in one of the families acknowledged her affection and respect by naming her own daughter after Deborah (Doll, 1983).

In the early 1920s, when an epidemic broke out in one of the buildings at the state school, Deborah served as a nurse's aide. It was reported that she "mastered the details of routine treatment and was devoted to her charges." During this period, a patient bit Deborah's hand as she was feeding her. One of her fingers was so badly injured that it later had to be amputated. According to Helen Reeves (1938), Deborah wore "this disfigurement as a badge of honor" (p. 196).

On occasion, Deborah accompanied the official families that she worked for to the shore for holidays. Her preference in vacations, however, seems to have been for a series of yearly excursions that she and social worker Reeves took together. Here is Reeves's recollection of their 1939 autumn trip to Washington, D.C.:

> As we rolled along southward I did not realize—though I should have—that I was establishing a precedent and that the succeeding five years would find me doing exactly the same sort of thing at this season of the year. Nineteen-forty would see us at the World's Fair in New York City; Luray Caverns would be visited in 1941 and Niagara Falls the year following; New York City again in 1943, and then—gasoline being scarce and travel facilities constricted—1944 would find us in Philadelphia for those three precious days. (Reeves, 1945, p. 3)

One of the photographs of Deborah in Goddard's book shows her sitting with a cat on her lap. Apparently she raised a long line of Persians and particularly relished the kittenhood of her charges, constructing a pink and blue bassinet to shelter the new arrivals. Her kittens were popular in the institution, and she sold them to a select clientele of training school employees at bargain prices. Her cat family grew faster than the market for them, however, and she was eventually forced to keep only one, her favorite, Henry. "He is named for a dear, wonderful friend who wrote a book. It's the book what made me famous" (Reeves, 1938, p. 194).

Deborah had a love of nature that she expressed in many ways. Eugene Doll writes:

> Her published photographs show her with stray animals she had befriended; unpublished ones show her peeking coyly through the apertures of a rose garden. In the spring she loved to walk among the daffodils and flowering shrubs. "She had a child's appreciation for the daisies and the dandelions or a bouquet of colorful leaves." She was fond of church and religious festivals,

alternately exulting and suffering on Christmas and Good Friday. She reveled in the rhythm of poetry. (Doll, 1983, pp. 31–32)

On her excursions with Helen Reeves, Deborah kept bits of toast from breakfast in her handbag on the chance that they might encounter a bird or squirrel. She loved visiting Central Park, the Museum of Natural History, and the Bronx and Philadelphia Zoos. According to Reeves, Deborah considered her devotion to animals her greatest virtue (Reeves, 1945, pp. 6–7).

Deborah's beauty is evident from photographs of her in the Kallikak book. Her charm and attractiveness are frequently mentioned by those who knew her. The photograph in Illustration 5 shows her working as a waitress at the training school.

Doll quotes one acquaintance as saying, "Hers was a body which moved with full knowledge of the impact of its movements on the opposite sex." He goes on to cite the impression of an employee who had accompanied a group of the institution's girls on a boardwalk stroll: "Everytime we passed a man or group of men, they would stop, turn, look after Deborah, and occasionally start to follow us. I do not know what signals Deborah was sending out, but it seemed that one glance from her eyes could summon a following. I was uneasy until we got home, though Deborah had done nothing really fresh or out of order " (Doll, 1983, p. 32).

While Deborah was serving as a nurse's aide during the epidemic, she stayed in a room near the patients. There she was not under the same close supervision of her usual living area. It appears that her woodworking skill enabled her to alter her window screen for easy exit and entry. She had fallen in love with an employee of the state school (apparently a maintenance worker). They seemingly enjoyed the moonlit grounds and each other in a romantic interlude before being discovered. The young man was "kindly dismissed by a lenient justice-of-the-peace" and regulations

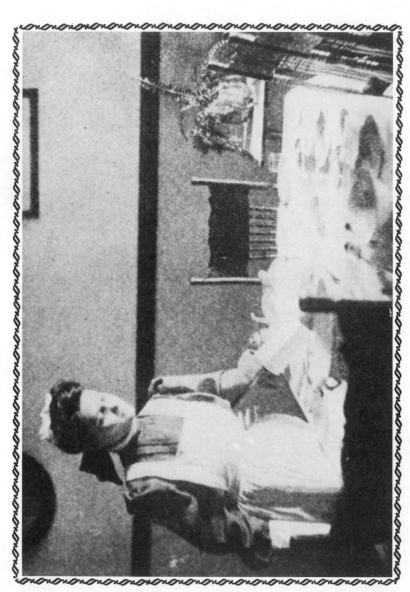

Illustration 5: Deborah Kallikak as a waitress at the Vineland Training School. Reprinted from The Kallikak Family: A Study in the Heredity of Feeble-Mindedness by H.H. Goddard, 1912, New York: Macmillan.

were tightened for Deborah (Reeves, 1938, p. 196). After a similar experience sometime later, Deborah mourned, "It isn't as if I'd done anything really wrong. It was only nature!" (Reeves, 1938, p. 197). Years afterward, she would again fall in love. Helen Reeves gives us some insight into the institutional attitude concerning Deborah's feelings of love and her right to romantic involvement.

> In the early fall of 1939 I returned to Vineland after a month's leave to find Deborah's spirits and morale at low ebb. She had worked hard during the summer, trying to do justice to a housework job for one of the official family, keeping on meanwhile with her responsibilities as custodian of the gymnasium and costume room. She had also managed to fall in love while I was away, which romance had been discovered and quietly nipped in full bloom without her knowledge. (Reeves, 1945, pp. 2–3)

How can it be that a woman of considerable talent in several areas of her life, a woman of beauty and charm, a woman lacking in academic skills but able to perform productive work is institutionalized for 81 of the 89 years of her life? When so much of the information that is available indicates that Deborah had the potential for living in society, what factors contributed to her lifetime of segregation?

Repeatedly in the accounts of Deborah's life, references are made to her appearance of normality. Visitors and new employees often expressed disbelief when told that she was mentally retarded. Time and again, such skepticism about the validity of classifying Deborah as feebleminded, as a moron, was countered with the results of standardized intelligence tests. Throughout the available reports, her performance on tests of academic or abstract ability was held to be of greater importance than the obvious strengths she demonstrated in her daily life. All subsequent descriptions echo to some degree Goddard's summation of Deborah's condition:

This is a typical illustration of the mentality of a high-grade feeble-minded person, the moron, the delinquent, the kind of girl or woman that fills our reformatories. They are wayward, they get into all sorts of trouble and difficulties, sexually and otherwise, and yet we have been accustomed to account for their defects on the basis of viciousness, environment, or ignorance.

It is also the history of the same type of girl in the public school. Rather good-looking, bright in appearance, with many attractive ways, the teacher clings to the hope, indeed insists that such a girl will come out all right. Our work with Deborah convinces us that such hopes are delusions.

Here is a child who has been most carefully guarded. She has been persistently trained since she was eight years old, and yet nothing has been accomplished in the direction of higher intelligence or general education. Today if this young woman were to leave the Institution, she would at once become a prey to the designs of evil men or evil women and would lead a life that would be vicious, immoral, or criminal, though because of her mentality she herself would not be responsible. There is nothing that she might not be led into, because she has no power of control, and all her instincts and appetites are in the direction that would lead to vice. (Goddard, 1912, pp. 11–12)

Goddard eventually tempered his thinking on the issue of the unmodifiable nature of feeblemindedness, on the incurability of the moron. Deborah, however, would be affected by the legacy of the original diagnosis for the rest of her life. Perhaps the greatest tragedy was that Deborah came to believe that life in an institution was the only one possible for her. In 1938, she told Helen Reeves, "I guess after all I'm where I belong, I don't like this feeble-minded part but anyhow I'm not i-idic like some of the poor things you see around here" (Reeves, 1938, p. 199). In 1945, Reeves reported that "Deborah, in spite of her conscious superiority, does not feel secure away from the institution. . . . 'The world is a dangerous place,' she will tell you" (p. 2).

Deborah was confined to a wheelchair during her final years. She was often in intense pain because of severe arthritis and was unable to continue with the crafts that she had loved so much throughout her life. In these last years, she was offered the alternative of leaving the institution to live in the community from which she had been segregated for almost all of her life. She declined the opportunity; she knew that she needed constant medical attention (Personal communication, April 5, 1979, from H. Schultz, director of the Vineland State School, cited in Scheerenberger, 1983). Surely the outside world must have appeared by then to be a dangerous place to be; for her, the institution was the only community she understood and trusted. "As long as she was able she sent her friends photographs and dictated letters (she could no longer write) of the meaningful events of her life. Not only did she pride herself on her fame, she made a profound impression on all who knew her, and had a queen's knack for inspiring devotion" (Doll, 1983, p. 32).

In the Kallikak book, Goddard describes the custom of the children at the training school writing letters to Santa Claus about their Christmas wishes. He lists Deborah's requests from age 10 through 22 (Goddard, 1912, pp. 8–9). Although his reason for including it in the book is not clear, the list of requests both illustrates Deborah's development during those years and provides a poignant summation of her life.

- 1899—book and harmonica
- 1900—book, comb, paints, and doll
- 1901—book, mittens, toy piano, handkerchief, and slate pencil
- 1902—wax doll, ribbon, and music box
- 1903—postcards, colored ribbons, gloves, and shears

- 1904—trunk, music box, fairy tales, games, ribbons, and big doll
- 1905—ribbons of different colors, games, hand-kerchiefs, music box, and fairy tales
- 1906—pair of stockings, ribbons, and rubbers
- 1907—watch, red ribbon, brush and comb, and paper
- 1908—three yards of lawn (light cotton fabric) and rubbers
- 1909—nice shoes and pink, dark blue, and white ribbons
- 1910—money for dentist bill
- 1911—rubbers, three shirts, blue scarf, three yards of linen, and two yards lawn for fancy work

NOTES

1. The J.E. Wallace Wallin cited in this extract was a psychologist and educator who had worked at the Vineland Training School early in his career and returned often for visits. He was a pioneer in the development of public school programs for mentally retarded children.

REFERENCES

Doll, E.E. (1983). Deborah Kallikak: 1889–1978, a memorial. *Mental Retardation, 21*, 30–32.

Goddard, H.H. (1912). *The Kallikak family: A study in the heredity of feeble-mindedness.* New York: MacMillan.

Reeves, H.T. (1938). The later years of a noted mental defective. *Journal of Psycho-Asthenics, 43*, 194–200.

Reeves, H.T. (1945). Travels with a celebrity. *Training School Bulletin, 42*, 1–9.

Scheerenberger, R.C. (1983). *A history of mental retardation.* Baltimore: Brookes.

Chapter 4

The Context

IN SEPTEMBER OF 1887, REVEREND S. OLIN GARRISON opened a school in his home in Millville, New Jersey. Its mission was the education of retarded children. Garrison was following in the footsteps of his father who, as representative from Cumberland County in the New Jersey State Legislature had, in 1842, introduced a bill that would have created a state home for feebleminded children; the measure was not passed (Myers, 1945). After its opening, Reverend Garrison's school was in such demand that he was soon searching for a larger facility. Benjamin Maxham of Vineland offered Garrison his home, the Scarborough mansion, and forty acres of land for the school. The new school was occupied in March, 1888, by the Garrison family and seven students (Crissey, 1982; Devery, 1939).

By 1898, the school had grown so large that Garrison needed an assistant. That year he hired Edward R. Johnstone as vice principal. Before coming to Vineland, Johnstone, formerly a secondary school teacher and principal, was director of the education department of the Indiana State School for the Feeble-Minded. Following Garrison's death, Johnstone was appointed principal in 1901. Johnstone was skilled at enlisting support for the school from wealthy and powerful persons. With their assistance, the school grew in size and reputation under his leadership (Crissey, 1983).

Central to Johnstone's efforts was his belief that the feebleminded should be sequestered from society and cared for humanely. Many people advocated stopping the propagation of feeblemindedness by enforced sterilization. To some of its advocates, this solution would render institutions unnecessary and allow the feebleminded to remain in society. Johnstone rejected this idea. His opposition was based on the belief that sterilization would only encourage sexual vice. He was also opposed to marriage among feebleminded people under any circumstances. His approach was to segregate those who were deemed feebleminded, to prevent their marriage and reproduction, and to provide them with kind treatment (Voorhees, 1981).

In 1906, Johnstone established a research department at Vineland. Its purpose was to investigate the causes of mental retardation and possible means of its prevention. Before establishing the department, Johnstone asked G. Stanley Hall, the pioneer American psychologist and president of Clark University, to recommend someone capable of directing the proposed research. Hall suggested his former doctoral student, Henry Herbert Goddard (Crissey, 1983). Johnstone was already acquainted with Goddard, who had served as a consultant to the training school while teaching psychology at the Pennsylvania State Normal School. Johnstone acted on Hall's advice and hired Goddard, and the research at Vineland began (Voorhees, 1981).

Henry Goddard was born in Vassalboro, Maine, in 1866. His father died when he was very young, and he was supported from his mother's earnings as a traveling Quaker minister. The Quaker church (Society of Friends) also provided additional assistance to the family over the years. Goddard received his bachelor's degree from Haverford College in 1887. After teaching for one year at the University of Southern California, he returned to Haverford and earned a master's degree in 1889. That same year he married Emma Florence Robbins. They were to be childless. Until 1896, Goddard served as a secondary school principal. He then entered Clark University where he studied psychology with Hall. He received his Ph.D. in 1899. It was then that he joined the faculty of the Pennsylvania State Normal School in West Chester (Voorhees, 1981). A photograph of Goddard after his appointment at the Vineland Training School is shown in Illustration 6.

Goddard evidently wrote to Hall shortly after his appointment at Vineland, asking for ideas about the direction that the research there should take. In his reply, Hall said, "In response to your favor of the 2nd, would say that I have no doubt psychological study of defectives could be made valuable to science as well as to the institution" (Hall, 1906, p. 16). He offered no suggestions for the overall direction of the research. However, he did list areas that might be of interest in the study of retarded children, such as vocabulary, dress, musical abilities, and play.

Goddard also corresponded with Adolph Meyer, then director of the Pathological Institute of the State of New York.[1] To Goddard's request for advice, Meyer responded with some suggestions for keeping records and test information on each child. However, he cautioned that "the chief thing to guard against is to pile up a lot of apparently very scientific tests which in the eyes of the teacher and any common-sense individual would appear to be top heavy, and therefore bring discredit to the movement" (Meyer, 1906, p. 16).

This, however, was not the kind of advice Goddard was seeking. Rather than suggestions concerning which behaviors might be interesting to observe or how to keep records, he was looking for a comprehensive theoretical base on

Illustration 6: Henry Goddard circa 1912. Reprinted with permission from the Archives of the History of American Psychology, Bierce Library, University of Akron.

which to design and conduct his research. He was later to express frustration that he was unable to find assistance in this quest (Goddard, undated manuscript).

Goddard spent his first years at Vineland observing and interacting with the children. He also conducted tests with physical and behavioral measures similar to those that Hall had suggested. None of these investigations, however, seemed to lead him any closer to an understanding of the nature and causes of mental retardation. Apparently discouraged, Goddard decided to go abroad to see what the Europeans were doing in the study and treatment of feeblemindedness (Crissey, 1983).

While traveling in Europe, Goddard met Alfred Binet and Theodore Simon and became interested in their work on intelligence testing. Upon his return to the United States, he had the Binet test translated into English under the auspices of the Vineland research laboratory. After trying the test on residents of the training school, he became convinced that it was a valid and powerful instrument. He was confident of the test's ability to detect the presence and degree of feeblemindedness. The Binet test would become the method through which the subtle retardation of the moron could be revealed—a cornerstone of Goddard's research.

At about the same time, Lewis Terman, another of Hall's former students and a friend of Goddard, was working on a revision of the Binet test in California. His version would become widely adopted and be used in clinical and research settings. It was Goddard, however, who first used the test to classify mentally retarded people—that is, as idiots, imbeciles, or morons—and to establish the need for institutionalization. He would also use the tests to detect feeblemindedness in school children, delinquents, and immigrants. Results from the Binet test were presented as proof that Deborah Kallikak was a moron and were used to justify her placement in the training school.

A second foundation for his research was found by Goddard closer to home. The Carnegie Foundation in Washington, D.C., had received a donation from Mrs. E.H. Harriman in New York to establish a center for the study of human eugenics. Mrs. Harriman, widow of the chairman of the Union Pacific Railroad and mother of W. Averell Harriman who later served as Under Secretary of State and governor of New York, had become convinced that hereditary defects posed a threat to society. She eventually contributed more than a half million dollars to the center. She also provided the primary financial support for the Committee on Provision for the Feeble-Minded, a national outgrowth of an extension program at Vineland designed to educate the public concerning the menace of feeblemindedness and the need for more institutions in which to contain people found to be defective (Haller, 1963).

A primary aim of the Carnegie center was to conduct research on the inheritance of defective traits and the prevention of their transmission. The Eugenics Record Office was established at Cold Spring Harbor on Long Island, and Charles B. Davenport was appointed its director. Davenport was a biologist with a zealous belief in the hereditary basis of most of humanity's physical, mental, and social ills. He was enthusiastic about his campaign to eliminate these ills by controlling human reproduction.

According to psychologist Marie Skodak Crissey:

The Eugenics Record Office of the Cold Spring Harbor had as its objective the "accumulation and study of records of physical and mental characteristics of human families to the end that people may be better advised as to fit and unfit marriages." Among those on the advisory committee were Alexander Graham Bell, now best remembered for the telephone, and Luther Burbank for his perfection of the potato.

Since both the Cold Spring Harbor and the Vineland Laboratory under Goddard had common interests, it is not surprising to find that Bulletin No. 2 of the Eugenics

Record Office is devoted to "The Study of Human Heredity: Methods of Collecting, Charting, and Analyzing
Data," with Goddard and Johnstone as principal
authors. There were frequent contacts between Vineland and Cold Spring Harbor in the 1907–1917 period
since their interests and philosophies overlapped.
(Crissey, 1983, p. 60)[2]

The techniques used by Goddard in the Kallikak study
and in subsequent research were surely influenced by the
methods developed by the Eugenics Record Office. Even
more important, however, was the influence of the eugenics
philosophy, derived from social Darwinism, which started
him on a search for evidence that feeblemindedness is
largely hereditary. Eugenics, according to Charles Davenport, was the "science of the improvement of the human race
by better breeding" (Davenport, 1911, p. 1). The aim of the
eugenics movement was to conduct hereditary research
that would result in the upgrading of the human stock,
similar to the way genetics was being applied in agriculture
and animal husbandry. People with superior traits were to
be encouraged to reproduce early and often. People with
defective characteristics were to be prohibited from
reproduction. To achieve these ends, however, the eugenicists needed evidence that the undesirable traits they
wanted to eliminate were in fact hereditary. Compelled by
eugenic philosophy, Goddard came to believe that he had
found such evidence embodied in the Kallikak family.

In June of 1908, Goddard submitted a proposal to the
Carnegie Institute, asking for $25,000 to support his research
at Vineland. One component of the proposal called for field
research on the hereditary background of children at the
training school: "Full history can only be obtained by the
visit of an expert to the home and neighborhood." He asked
for $5,000 for salaries and traveling expenses for field work-
ers, "a man and a woman with assistants to collect data on
heredity" (Goddard, 1908, p. 16).

I have found no records to indicate that the Carnegie Institute ever awarded funds for the heredity research. There are statements in several places, however, indicating that the research at the training school was financed by private donors. Much, if not most, of the research was under-written by Samuel S. Fels of Philadelphia.

In a letter to Goddard, Fels, seeking to clarify what was apparently a misunderstanding concerning the extent of his backing of the research, stated:

> I am in receipt of your favor of the 29th, and there seems to be a misunderstanding.
>
> My recollection of the matter is, that I sent you a check and told you, that if you got hard up again, I would be glad to help you out, but have no remembrance of saying at any time that I would take care of all the finances for this special work.
>
> Some time ago you told me that there has been other amounts of money received. How much they were and from whom they were has escaped me. . . .
>
> Do not let this difference of understanding worry you, as it is not bothering me. After we get in touch, the matter will be cleared up very easily. (Fels, 1910, p. 51)

Goddard was very prompt in responding and in his reply explained:

> I remember that the first time you intimated that you wished to contribute you named $1,000 as the amount you would try and see how it came out. We then agreed that we should pay Miss Bell and Miss Hill their salaries for a year and the balance would start a field worker. In the Fall Mr. Van Wagenen gave $500 and we started a second field worker . . . it was then that I understood you to say that you thought you would like to be respon-sible . . . for the hereditary work and you said you thought we better get another field worker and possibly

later a fourth . . . I found one in about two weeks.
(Goddard, 1910, p. 51)

Samuel Fels was president of Fels and Company, the
manufacturer of Naptha soap. His parents were Jewish immigrants who had left Germany after the Revolution of 1848.
The family struggled hard, and their soap business was
eventually one of the most successful of its kind. Eventually,
Mr. Fels became best known as a civic leader and philan-
thropist. He reportedly donated more than 40 million dollars
to various projects and causes during his lifetime. His phi-
lanthropies included the Research Institute of Temple University Medical School and the Research Institute of Human
Development at Antioch College (Voorhees, 1981).

Samuel Fels helped organize the Hebrew Immigrant Aid
Society in 1884, the Federation of Jewish Charities in 1901,
and the Allied Jewish Appeal in 1938 (Phalen, 1969). It is
ironic, given Fels's support of these causes and the fact that
he was Jewish himself, that Goddard's research would
extend to intelligence testing of immigrant groups and result
in the conclusion that a large majority of Jewish immigrants
were morons. Reports of this finding contributed to the successful efforts to pass the Immigration Restriction Act of 1924,
which practically eliminated Jewish immigration to the
United States.

The dedication of the Kallikak book reads as follows:

To
Mr. Samuel S. Fels
A Layman With The Scientist's Love of Truth
And The
True Citizen's Love Of Humanity Who Made
Possible This Study And Who Has Followed
The Work From Its Incipiency With
Kindly Criticism And Advice
This Book Is Dedicated

(Goddard, 1912)

NOTES

1. Adolph Meyer taught and directed a clinic at Clark University while Goddard was studying there. He became chairman of the department of psychiatry at Johns Hopkins Medical School in 1914.
2. With respect to the inclusion of Alexander Graham Bell on the Eugenics Record Office's advisory committee, it is of interest to note that Bell had studied hereditary deafness and was concerned that intermarriage among the deaf could lead to a "deaf variety of the human race" (Haller, 1963, p. 32).

REFERENCES

Crissey, M.S. (1982). There was a little school house. *Education and Training of the Mentally Retarded, 17,* 305–306.

Crissey, M.S. (1983). The searchlight of science. *Education and Training of the Mentally Retarded, 18,* 59–61.

Davenport, C.B. (1911). *Heredity in relation to eugenics.* New York: Henry Holt and Company.

Devery, E.D. (1939). *The story of Four Mile Colony: A successful demonstration of human conservation.* Vineland, N.J.: New Jersey Department of Institutions and Agencies.

Fels, S.S. (1910). [Letter]. *Goddard papers* (Box M35.2, Miscellaneous). Akron, Ohio: University of Akron, Bierce Library, Archives of the History of American Psychology.

Goddard, H.H. (undated). [Manuscript]. *Goddard papers* (Box M37, E). Akron, Ohio: University of Akron, Bierce Library, Archives of the History of American Psychology.

Goddard, H.H. (1908). [Proposal]. *Goddard papers* (Box M33, No. 2, p. 16). Akron, Ohio: University of Akron, Bierce Library, Archives of the History of American Psychology.

Goddard, H.H. (1910). [Letter]. *Goddard papers* (Box 35.2, Miscellaneous). Akron, Ohio: University of Akron, Bierce Library, Archives of the History of American Psychology.

Goddard, H.H. (1912). *The Kallikak family: A study in the heredity of feeble-mindedness.* New York: Macmillan.

Hall, G.S. (1906). [Letter]. *Goddard papers* (Box M33, No. 2, p. 16). Akron, Ohio: University of Akron, Bierce Library, Archives of the History of American Psychology.

Haller, M.H. (1963). *Eugenics: Hereditarian attitudes in American thought.* New Brunswick, N.J.: Rutgers University Press.

Meyer, A. (1906). [Letter]. *Goddard papers* (Box M33, No. 2, p. 16). Akron, Ohio: University of Akron, Bierce Library, Archives of the History of American Psychology.

Meyers, W.S. (1945). *The story of New Jersey.* New York: Lewis Historical Publishing Co.

Phalen, D. (1969). *Samuel Fels of Philadelphia.* Philadelphia: Samuel E. Folo Fund.

Voorhees, D.W. (Ed.) (1981). *Dictionary of American Biography.* New York: Charles Scribner's Sons.

Chapter 5

Elizabeth Kite

In the preface to the Kallikak book, Goddard pays tribute to his field workers:

> I wish also to make special mention of the indefatigable industry, wisdom, tact and judgment of our field workers who have gathered these facts and whose results, always continually checked up, have stood every test put upon them as to their accuracy and value.
>
> The work on this particular family has been done by Elizabeth S. Kite, to whom I am also indebted to practically all of Chapter IV. (Goddard, 1912, pp. x–xi)

Elizabeth Kite was born in Philadelphia in 1864. Her parents were Quakers, and she was brought up in a somewhat conservative atmosphere. She received her secondary education in a Quaker boarding school and subsequently took university courses in England, France, Germany, and Switzerland. In 1906, while studying in England, she converted to Catholicism. It is interesting to note that many members of her family, including her father, who was a Quaker preacher, followed her into the Catholic Church.

Prior to coming to Vineland, she had served as principal of a private school in Philadelphia and had taught science in California. She had also taught botany, French, and German in a school in Nantucket, Massachusetts (Hoehn, 1981).

The earliest evidence that I have been able to find of Elizabeth Kite's association with Goddard and the Vineland Training School is a letter from her to Goddard, dated February 28, 1910. In the letter, she indicates that she is living in a home owned by Samuel Fels. She discusses a recent visit by Mr. and Mrs. Fels. During the visit, plans were made to use the house as a "fresh air" retreat of sorts for city children (Kite, 1910). She apparently had a longstanding relationship with the Fels family, and it is likely that she was hired at Vineland on the suggestion of Samuel Fels. In a letter to Fels later in 1910, Goddard says, "I told Miss Kite that inasmuch as you were financing the hereditary work I would consult you before advancing her wages permanently" (Goddard, 1910, p. 51).

Kite's 1910 letter indicates that she had been working on the general heredity project but had probably not yet begun to focus on the Kallikak family:

I am hoping to finish up that Bridgton work sometime in the next two weeks, and come back by way of Vineland to get together the rest of the Camden County work, perhaps I shall have the opportunity of seeing Prof. Johnstone then. . . . I am enclosing last months expenses, also an account of my Milleville and

Leesburg experiences with their results. (Kite, 1910, p. 51)

Elizabeth Kite seems to have approached her work with zeal and a spirit of self-sacrifice. In response to a letter from Goddard offering to raise her salary, she writes:

> It will surprise you to know that your letter made me very angry . . . I fixed the rate at which I could work at the lowest possible figure at which I could leave home and I should despise myself forever if I accepted an increase of seventy-five cents a day. I do beg you Dr. Goddard not to be hurt by the frankness of my letter . . . if I have been able to serve you to your satisfaction I am more than repaid—besides I have learned myself far more than you have gained from me, so you need feel no debt of obligation. (Kite, undated letter, p. 51)

In the Archives of the History of American Psychology at the University of Akron I found a manuscript written by Elizabeth Kite as her personal account of the Kallikak study. I have found no evidence that it was ever published. Its considerable importance and interest in connection with the Kallikak study, however, warrant inclusion of substantial portions of the Kite manuscript at this juncture.

After explaining that the Kallikak investigation had begun as part of the hereditary study of residents of the training school and following a brief description of Deborah, Kite discusses the initial steps she took in constructing the Kallikak genealogy:

> It was from this point that the investigation started. Through the charitable woman before mentioned it was possible to locate the family, make the acquaintance of its different members and study their mentality. In time the investigation was extended to the brothers and sisters, aunts, uncles and cousins of Deborah's mother.

This work was made possible by the fact that they were found scattered through a prosperous rural community where the grandfather of Deborah had lived for many years, working about as a farm hand. Many residents of the section remembered the old man and his eleven children well and could point out the shack where they had once lived. It was learned also that Deborah's grandfather was one of twenty brothers and sisters. These were in time all sought out and when living and where near enough were visited and their families studied. Gradually then our acquaintance grew until a complete record was obtained of three generations.

It proved however a very difficult matter to get farther back than Deborah's grandfather. Who was the father of this man and of his nineteen brothers and sisters? Where did they come from? A rumor was frequently encountered that told of their coming from a mountain ridge farther up the State but the information was too vague to build upon until a happy coincidence led me to the acquaintance of a charming elderly lady whose girlhood days had been spent in a town of some importance situated at the foot of the aforementioned ridge. "Why yes, I knew Justin Kallikak—he used to work for my father after we left our mountain town. It was his father who came from the same place we did only they lived back in the woods. They were the funniest people. When I was a little girl I remember being taken to drive to see the old hut, with window frames stuffed with rags, where they then lived. The old man—grandfather of Justin—always went by the name of 'the old Horror'—he was so greasy and fat. On election day he'd come into town, dressed in somebody's old clothes given him for the occasion, so he'd vote their ticket. I remember his daughters too—Old Sal—Old Moll—and Jemimah— who lived with him. There were dreadful scandals about them but I have forgotten the details. Jemimah used to come into town selling huckleberries—a great, tall, angular creature, in men's boots, short calico dress and slab sunbonnet."

Later a sister of Justin, who lived near Deborah's mother, an old woman of over eighty with whom I became very friendly, gave me the name of a cousin of her's who lived back in the mountains. Armed at last with a name and address it was but a work of time and patience to complete the genealogy back to "the old Horror" who turned out to be Martin Kallikak, Jr." (Kite, undated manuscript, pp. 3–4)

Kite then goes on to discuss her discovery of the two Kallikak lines and her search for a connection:

From the beginning of the investigation it was apparent that an eminently respectable line of Kallikaks was existing in New Jersey and surrounding states, in many cases side by side with ancestors or remote relatives of our "Deborah," although a veritable gulf separated them from one another, socially as well as intellectually.

It was early learned that a Genealogy existed of the family and the statement was made by an intelligent representative "You'll not find a Kallikak anywhere in the United States or Canada who is not descended from Casper, our ancestor who came to New Jersey the latter part of the 17th century." This led to the hope that evidence might be found for connecting the two lines and this in spite of their social divergence. . . .

But who was Martin Kallikak Jr.? Who was his father? What of his brothers and sisters? Or was he an only child? Regarding these important questions no important information was obtainable. The time was too remote for living persons to remember and only vague traditions were to be found floating in the family. As for documents there were none that could be made to bear upon the subject although graveyards, county registers of marriages, births and deaths were thoroughly gone over. The case seemed hopeless. . . . One day early in the Spring the lure of the unsolved problem led me back into the hills. Who was the father of Martin, Jr.? Was it true he had no brothers, these and similar questions were constantly ringing in my ears. Intent upon finding their solution, I stopped at the cabin of Mary Ann whom I had not seen for many months. The old woman was just recovering from a long illness and was so happy to see a visitor that she was more talkative than usual. Always in the past when conversation went back to her grandfather's relatives, she became silent and it had been impossible to gain any information. This was attributed to ignorance of the facts, something not at all surprising. This time, however, she was in a reminiscent mood and talked freely of old times. When the question was again put—"By the way, did you tell me once that your grandfather had no brothers?" A strange look came into her face and she glanced up shyly as she said "he had a half-brother." "Oh," I said, "his mother was married twice?" Then quickly but still shyly she answered, "no, ye see she had him *before* she was married." "And she named him?" I asked eagerly for my mind was beginning to grasp the real significance of the disclosure she had made. "Yes, she named him after his father." "Oh," I gasped, and then realizing that an indiscreet show of interest might arouse her suspicion and prevent further disclosures, I rapidly regained self control and listened with an outward indifference, but with an inward emotion that set my heart beating sledge hammer blows, to the rest of her story. "Ye see," she went on, "Frederick Kallikak who lived about twenty miles from us was his half-brother. Course he never noticed my grandfather, but they looked alike. If they'd a been dressed alike you'd a thought they was twins." (Kite, undated manuscript, pp. 3, 8–9)

Elizabeth Kite concludes her story with an account of how she obtained confirmation of Martin Kallikak, Sr.'s, paternity of the "bad" line. She visited the town where Martin, Jr.'s, mother had lived and talked with an elderly resident of the area who told her:

> "Why do you know I haven't thought of those people for years, but as I talk to you the memories keep coming back. Old Martin was himself an own half-brother to Frederick Kallikak who was a fine man and a gentleman and who lived about twenty miles from here on a splendid farm that his father left him." "Their father was the same then," I asked. "Yes that's the way it was," "And his mother," I went on. "I never saw her but I've heard about her when I was a boy. She lived with an old soldier named_____. He was very queer. Of course, they weren't married. They lived on the edge of the town over there. I can point out the place where the old house stood." (Kite, undated manuscript, p. 11)

Kite's final comment in the manuscript is a statement of her belief that the manuscript's description of her research should be adequate evidence of its validity. She felt that the work rested upon the words of established witnesses.

And so it becomes very clear that Goddard relied completely on Elizabeth Kite's collection and interpretation of data as the basis of the Kallikak study. He took the genealogical information she had gathered, organized it, and then added a Mendelian explanation for the differential status of the two branches of the family.[1]

His reliance on Kite is strikingly illustrated in a letter he wrote in 1928 when the Kallikak study was facing increasing criticism. He inquired of Kite:

> Did we ever know the real name of the mother of the bad line in the Kallikak story? The one that I called the nameless girl.

One or two people, including Porteus, who are opposed to the idea of the heredity of feebleminded-ness, have attempted to discredit the story of the Kallikak family, among other things stating that it is absurd to attempt to declare that this girl was feeble-minded when so little is known of her that we do not even know her name. I should like to turn the tables on them if possible by stating that we did know her name and that calling her "The nameless feeble-minded girl" was in accordance with our policy of disguising *all* names. (Goddard, 1928, p. 35)[2]

In her reply, Elizabeth Kite supplied some of the same information that was included in her previously cited man-uscript. She could not provide a name for Martin, Jr.'s, mother, but she does quote a man she interviewed as saying that she was "not all there you know." (Kite, 1928, p. 35)

The years following the publication of the Kallikak study were busy ones for Miss Kite. Drawing upon her fluency in French, which she had developed while studying in Europe, she translated *The Development of Intelligence in Children* and *The Intelligence of the Feebleminded*, both by Alfred Binet and Theodore Simon. These were pioneering works, and their translation into English did much to facilitate the acceptance of intelligence testing in the United States. It is likely that the translation of the *Binet-Simon Measuring Scale for Intelligence*, first published in English by the Vine-land Training School, was also done largely by Elizabeth Kite.

In 1913, she had an article, "The Pineys," published in a popular social science periodical. It was based on her expe-riences with families she encountered in the Pine Barrens of New Jersey while conducting hereditary research for the training school. The article is reminiscent of the Kallikak study, complete with genealogical charts, family pictures, and a discussion of the social implications of familial trans-mission of mental retardation (Kite, 1913).

Kite's article was reported widely in the popular press. Many newspapers printed excerpts. All over New Jersey people reacted with alarm to the reported conditions in the Pine Barrens. James T. Fiedler, then governor of the state, visited the area and, perhaps responding in a politically expedient fashion, recommended to the legislature that the Pine Barrens be somehow segregated from the rest of New Jersey in the interest of the health and safety of the people of the state. "I have been shocked at the conditions I have found," he said. "Evidently these people are a serious menace to the State of New Jersey because they produce so many persons that inevitably become public charges. They have inbred, and led lawless and scandalous lives, till they have become a race of imbeciles, criminals and defectives" (McPhee, 1967, p. 52).

In the same year her article appeared, Elizabeth Kite was appointed by the state of New Jersey to continue her investigations in the area and was provided funds to support the work. It was apparently the first appropriation ever made by a state legislature for such a purpose. One immediate result of her work was the establishment in 1914 of a new institution for the mentally retarded in the Pine Barrens area of the state (Devery, 1939).

The stigma generated by the study in the Pine Barrens became pervasive and persisted for many years. The derogatory implication of the term *Piney* was generalized to everybody in the region. This apparently surprised and appalled Miss Kite later in her life. In 1940 she told an interviewer, "Nothing would give me greater pleasure than to correct the idea that has unfortunately been given by the newspapers regarding the pines. Anybody who lived in the pines was a piney. I think it a most terrible calamity that the newspapers publicly took the term and gave it a degenerate sting. Those families who were not potential state cases did not interest me as far as my study was concerned. I have no language in which I can express my admiration for the pines and the people who live there" (McPhee, 1967, p. 54).

Illustration 7: Elizabeth Kite circa 1917. Reprinted with permission from the archives of the Department of Special Collections and Archives, Archibald Stevens Alexander Library, Rutgers University.

Elizabeth Kite was affiliated with the Vineland Training School until 1918, and she returned for a year in 1927 to update her studies in the Pine Barrens. A photograph of her as she appeared just before leaving the training school is shown in Illustration 7.

After leaving the training school, Elizabeth Kite devoted herself to the study of historical relationships between France and the United States. She was also appointed archivist of the American Catholic Historical Association. She was the first woman to be awarded an honorary doctorate from Villanova University, and the French government conferred upon her the Croix de Chevalier de la Legion d'honneur in recognition of her work. She died at age 89 and was buried in Philadelphia (Hoehn, 1981; In memoriam, 1954).

Upon the publication of the Kallikak book, Goddard signed copies for Miss Kite, writing:

> To Elizabeth Kite—without whose indefatigable labor the material in this book would never have been brought to light; and without whose skill and excellent judgment would not have been worth publishing even when collected ("In Memoriam," 1954, p. 202).

NOTES

1. Goddard applied to issues of human feeblemindedness Gregor Mendel's discoveries in his famous genetic research on peas. Goddard argued that intelligence was a unitary trait determined by a single gene. Normal intelligence resulted from a dominant gene; feeblemindedness from a recessive gene. A normal carrier of a recessive gene (a simplex) could produce feebleminded children by mating with another carrier or by mating with a feebleminded person (a duplex).

2. Stanley Porteus, an Australian, was director of research at Vineland Training School from 1919 to 1925. He then became professor of clinical psychology at the University of Hawaii. He was best known for his Maze Test, which was touted as a culture-free measure of intelligence.

REFERENCES

Devery, E.D. (1939). *The story of Four Mile Colony: A successful demonstration of human conservation.* Vineland, N.J.: New Jersey State Department of Institutions and Agencies.

Goddard, H.H. (1910). [Letter]. *Goddard papers* (Box 35.2, Miscellaneous). Akron, Ohio: University of Akron, Bierce Library, Archives of the History of American Psychology.

Goddard, H.H. (1912). *The Kallikak family: A study in the heredity of feeble-mindedness.* New York: Macmillan.

Goddard, H.H. (1928). [Letter]. *Goddard papers* (Box 35.1, Correspondence). Akron, Ohio: University of Akron, Bierce Library, Archives of the History of American Psychology.

Hoehn, M. (Ed.). (1981). *Catholic authors: Contemporary biographical sketches, 1930–47.* Detroit, Mich.: Gale Research.

In memoriam, Elizabeth S. Kite. (1954). *Training School Bulletin, 50,* 201–202.

Kite, E.S. (undated). [Letter]. *Goddard papers* (Box 35.2, Miscellaneous). Akron, Ohio: University of Akron, Bierce Library, Archives of the History of American Psychology.

Kite, E.S. (undated). [Manuscript]. *Goddard papers* (Box M614, Folder). Akron, Ohio: University of Akron, Bierce Library, Archives of the History of American Psychology.

Kite, E.S. (1910). [Letter]. *Goddard papers* (Box M35.2, Miscellaneous). Akron, Ohio: University of Akron, Bierce Library, Archives of the History of American Psychology.

Kite, E.S. (1913, October 4). The "Pineys." *Survey,* pp. xx.

Kite, E.S. (1928). [Letter]. *Goddard papers* (Box 35.1, Correspondence). Akron, Ohio: University of Akron, Bierce Library, Archives of the History of American Psychology.

McPhee, J. (1967). *The Pine Barrens.* New York: Farrar, Straus, and Giroux.

Chapter 6

Acclaim, Criticism, and Defense

THE STORY OF THE KALLIKAKS WAS GREETED WITH ACCLAIM and achieved great popularity. The book became a best seller and went through several printings. Goddard's presentation of feeblemindedness as a hereditary problem was received with great interest by the general public and with enthusiasm by proponents of the eugenics movement. A simplistic explanation that social ills like poverty, prostitution, crime, and alcoholism were the result of feeblemindedness—specifically the high-grade, moron type—was appealing to the spirit of the time. To improve society, the "menace of the feebleminded" must be recognized and controlled.

Shortly after its publication, the book was given very favorable reviews in a number of periodicals. The *Dial* proclaimed:

> Dr. H.H. Goddard's volume entitled "The Kallikak Family" (Macmillan) is a remarkable human document. It is a scientific study in human heredity, a convincing sociological essay, a contribution to the psychological bases of the social structure, a tragedy of incompetence, and a sermon with a shocking example as a text. With an endless patience sustained by a scientific insight into the value of principle and detail, the history of two branches of a family has been traced. . . . Dr. Goddard and his associates have added notably to our insight into its fundamental significance, and particularly by demonstrating that deficient mentality—the stigma of an unworthy stock—is the clue to the condition, and vice and crime and inefficiency and brutality its issues under present-day social stress. (1912, p. 247)

The *Independent* said:

> This is the most convincing of the sociological studies brought out by the eugenics movement. It would be hardly possible to devise in the laboratory experimental conditions better adapted to produce a clear and decisive influence of heredity; nor could there be a more impressive lesson of the far-reaching and never-ending injury done to society by a single sin. (1912, p. 704)

The *American Journal of Psychology* called the study a "find" and praised Goddard for having "the training which enables him to utilize the discovery to the utmost" (1913, pp. 290–291).

The popular appeal of the Kallikak story was perhaps best indicated when, in 1913, Goddard was approached concerning the dramatic rights to the book. In March of that year, Alice Kauser, a dramatist's agent on Broadway, wrote:

> I want to apply for the dramatic rights of your book "The Kallikak Family." Joseph Medill Patterson, who has written plays with ideas back of them, is very much interested in it and has asked me to ascertain if an arrangement could not be entered into by which he could make a play out of your book. Will you please be so good as to let me hear from you. (Kauser, 1913, p. 121)

In reply, Goddard said:

> I shall have to take the matter up with the authorities here for while I am the author of the book, the material was gathered here and the book really goes out as representing the work of the institution in this line. I am sure we should have to be assured that the play would be one that would carry the moral lessons which the book is intended to convey. We would not consent to its being dramatized for any other purpose. Now, whether this can be done and still make it attractive and a success, you will know much better than I. (Goddard, 1913a, p. 121)

Goddard later arranged to meet with Alice Kauser in New York to discuss the possibility of a play. Included in his correspondence with her is mention of Blecker Van Wagenen, the previously noted benefactor of Goddard's research, who later became a strong advocate of compulsory sterilization of people judged to have undesirable traits: "In the meanwhile it would expediate matters if you cared to interview Mr. Blecker Van Wagenen with Dodd, Mead, & Co., on Fourth Avenue. He is one of our trustees and would speak with authority from that side of the matter" (Goddard, 1913b, p. xx).

Van Wagenen met with Kauser and discussed Mr. Patterson's interest in the book and the arrangements that might be made. He proposed to Goddard two conditions, however, that should precede any formal agreement:

> First, that some leading representative of the good Kallikak Family should be informed of the proposition and express willingness to have the history and the fictitious names used in the play.
>
> Second, that Messrs. Macmillan & Co. should be told of the plan, in order that if they saw fit to raise any question of rights or opposition to it, they might have the opportunity to do so before you become committed. (Van Wagenen, 1913a, p. 119)

Apparently the negotiations with Mr. Patterson did not go well. In a subsequent letter, Van Wagenen explained to Goddard that the terms that had been proposed for use of the book were objected to as being too high. He suggested that a more moderate proposal be made. Van Wagenen reported that he had asked Alice Kauser to tell Mr. Patterson that "we shall be glad to receive any proposition from him which he thinks would be satisfactory" (Van Wagenen, 1913b, p. xx).

I have found no evidence that Patterson eventually adapted the book for the stage or that a play based on the Kallikaks was ever produced on Broadway. Perhaps an agreement was never reached. Interest in the dramatic properties of the story, however, did not die. In 1925, after Goddard had left the Vineland Training School, D.L. James of Kansas City, Missouri, contacted Superintendent Johnstone at Vineland, describing a play he had written based on the Kallikak study. He had titled it *The Seed*. He asked that Johnstone forward a copy of the play to Goddard (James, 1925).

The next year, James responded to a letter he had received from Goddard, who was then teaching at Ohio State University:

> I have forwarded your letter to Mr. Charles Hopkins, who is to produce the play (probably next fall) and I

know that he will find most interesting what you say about the acting ability of the feeble-minded group. . . .

When a date is set for the first performance you may be sure that I shall be only too glad to let you know. It would mean much to Mr. Hopkins and to all of us to feel that you were in the audience. . . .

In reading the play did any other name suggest itself to you which you thought would be better or more interesting than *The Seed*? (James, 1926, p. 21)[1]

Again, I find nothing that documents that *The Seed* was ever published or produced.

Those reviews that contained critical comments on the Kallikak book were mild in tone and tended to focus on minor procedural or interpretive matters rather than fundamental aspects of the study. The following portion of the review from *Popular Science Monthly* is a good example:

A comparison of the two lines of descent from Martin Kallikak certainly exhibits a dramatic contrast, but it is scarcely the natural experiment in true heredity which Dr. Goddard claims it to be. If, on the one hand, Martin Kallikak had left neglected illegitimate children, without taint of feeble-mindedness, it is not likely that they would have established prosperous lines of descent. On the contrary, they would probably have intermarried with the degenerate and feeble-minded. If, on the other hand, the feeble-minded son had been legitimate, he would have been properly cared for, and in all probability would have left no such descendants as came from the illegitimate and neglected child. (1913, p. 416)

The eminent Harvard professor of genetics, E.M. East, praised Goddard's hereditary work as it was reported in the 1912 Kallikak book and in the subsequent volume, *Feeble-Mindedness: Its Causes and Consequences*. Noting that the findings were consistent with Gregor Mendel's unit trait inheritance (that is, that feeblemindedness is inherited through a single recessive gene), East stated:

> Again, the results of unions between a feebleminded parent (nn) and a normal heterozygote, a carrier (Nn), or between two carriers (Nn × Nn), are remarkably in accord with theory. They are even more closely in accord with theory than Goddard makes out in his report, for he did not make the proper corrections when calculating the expected number of feeble-minded children. . . . When the proper mathematical corrections are made in such cases, by a simple and correct algebraic method, the correspondence between theoretical expectation and actual result is so good as to be almost suspicious. (East, 1927, pp. 104–105)

It was not until 1926 that the first serious questioning of the Kallikak study occurred. In his book, *The Inheritance of Mental Disease,* Abraham Meyerson expressed skepticism concerning the concepts and techniques employed in the research. He was particularly critical of the idea that field workers with minimal training and experience could accurately diagnose mental retardation, either directly in the people they met or through second-hand accounts and memories of relatives and neighbors. He observed that the validity of the Kallikak study rested upon the accuracy of the diagnoses of feeblemindedness and that these were questionable:

> Really, it seems utterly unnecessary to have laboratories, blood tests, clinical examinations, and to take four years in medical school plus hospital experience, etc., when a woman can as a result of a dozen or two lectures make all kinds of medical, surgical, and psychiatric diagnoses in an interview or by reading through a court record . . . the keystone of the arch of their results and laws is the field investigator and her surmises as to the

mental and physical state of the dead and the quick. (Meyerson, 1925, p. 64)

Meyerson's criticism seems to have stimulated others to examine the study and its significance more closely. Soon several critical commentaries were published. As noted in our analysis of Elizabeth Kite's role in the research, Goddard wrote to her in 1928, complaining that a few people were attemping to discredit the story of the Kallikak family and asking for information that might help him in defending the work. She responded by sending him some facts concerning how the families had been traced. In early 1929, Goddard again wrote to Kite, this time thanking her for her help:

> It is four years since Meyerson wrote his stuff about the Kallikak Family, and I have paid no attention to it, but now Porteus comes out with a similar flare and Conklin, of Oregon, in his book on "Abnormal Psychology" says that Meyerson cast doubt upon the hereditary character of feeble-mindedness. It is very disturbing to find men who pretend to be scientists resorting to such "babyish" tricks in order to maintain their position. Neither of them make any attempt to disprove the figures and statistics and logic of the complete study of the 300 feeble-minded children. I am going to take the matter up in a part of my paper at Battle Creek on Thursday, and can use to direct advantage some of the statements in your letter. (Goddard, 1929, p. 35)[2]

Clearly, Goddard was bothered by the criticism of his research. He did little at this point, however, to answer his critics. This would change, however, as skeptics of the validity of his work grew in number and influence.

In 1939, Amram Scheinfeld discussed the story of the Kallikak family in his book, *You and Heredity*. He emphasized that the comparison of the "very good Kallikaks" with the "very bad Kallikaks" rested almost completely on the assumption that the illegitimate child whom the nameless feebleminded girl chose to call Martin Kallikak, Jr., was in fact the son of Martin Kallikak, Sr. He asserted that no court would accept the facts presented in the book as evidence of Martin, Sr.'s, paternity of the child. He then went on to say that even if the accuracy of his paternity were accepted, another point in genetics intruded:

> Granted that "Old Horror" (Kallikak, Jr.) was a degenerate because of bad heredity (and there is as yet no evidence that "degeneracy" is inherited) by what gene mechanism did he become that way? No single dominant gene could produce any such complex condition, nor is there any known gene that can singly produce even feeble-mindedness. Recessive genes would have had to be involved. Which means that as such genes must come from both parents for the effect to assert itself, no matter how chock-full of "black" genes the feeble-minded mother was, the worthy Martin Kallikak, Sr., himself had to be carrying such genes if the condition of his presumptive son, "Old Horror" was due to heredity. And that would mean, in turn, that the "good" Kallikaks also received some of those "black" genes! (Scheinfeld, 1939, pp. 361–362)

Scheinfeld's analysis was followed in 1940 by an unrestrained critique of the study and its impact by Knight Dunlap in *The Scientific Monthly.* According to Dunlap:

> Some thirty years ago the Kallikak family was boosted into an unfavorable notoriety, and shortly became a great asset to propagandists for eugenical sterilization and other nostrums. Even in books written by psychologists who ought to know better, the Kallikaks skulk in the corners of the pages, and leap out upon unwary students. The fame of the family began with an anecdote perpetrated with incredible innocence by an eminent expert on "intelligence," and repeated with

astonishing solemnity by many after him. The anecdote concerned the unblest union of a Revolutionary soldier with a feeble-minded girl, from which sprang a long line of descendants who were feeble-minded and prone to epilepsy, alcoholism, prostitution and what have you. I have often told this story to classes, and waited to see how many students would raise the obvious question: How do you know the girl was feeble-minded? Did anybody test her and assign an I.Q.? What is the evidence? Of course, there is no evidence. The promoter of the legend inferred that the girl was feeble-minded because she had feeble-minded descendants. Then, from the assumption of her feeble-mindedness he inferred the fatal heredity of amentia. This procedure, of assuming the conclusion in the premises from which it is presumably drawn, is called by the logicians, "Begging the question."

The Kallikak phantasy has been laughed out of psychology . . . but the theories involved . . . still linger in popular superstitions, endorsed by many writers of supposedly scientific books, along with other popular beliefs about heredity, and do definite damage to young persons who take the theories seriously. Many of these young persons fear to marry, lest their children should be feeble-minded, since they think there have been feeble-minded persons in their families in past generations. (Dunlap, 1940, p. 221)

A letter to Goddard from J.M. McCallie, a former colleague at the Vineland Training School, following the appearance of Dunlap's article is interesting in its tone and is exemplary of the kind of support that Goddard would receive from friends and former students as criticism of the Kallikak study became more intense:

The Scientific Monthly published by the American Association for the Advancement of Science—September,

1940—contains an article in it by one Professor Knight
Dunlap, Professor of Psychology, University of California at Los Angeles, and who, so he assumes, knows so
much psychology that he looks down with disdain upon
all others who have worked in this field. Why, he even
speaks of the Kallikak family as a laughable "anecdote
perpetrated with incredible innocence by an eminent
expert on intelligence." He further states that the
"Kallikak family has been laughed out of psychology,"
and, in substance, he gives one to understand that
everything one used to know about heredity "ain't so."
Well, of course you have read this article and feel duly
humiliated. It does seem too bad when some people find
out what some other people thought they knew, was
wrong, that they do not know how to be polite to the
erring ones." (McCallie, 1941, p. 5)

Similarly, L.N. Yepsen sent to Goddard a carbon copy of a
letter he had written to Waldemer Kaempeffert, science editor of the *New York Times*, charging that his review of
Scheinfeld's criticism of the Kallikak study was not knowledgeable of the manner in which the research was conducted:

> Having been intimately associated with Dr. Goddard, I
> know the care with which this work was initiated and
> carried out.
>   We all know that we cannot measure intelligence per
> se but we can certainly measure what it does. We see it
> every day here as repeatedly families of poor stock
> contribute to our welfare problem. (Yepsen, 1944, p. 96)

In a letter accompanying the carbon copy, Yepsen complained to Goddard that "Kaempeffert replied 'Thank you
very much!' That is all. Nothing more appeared in his column, of course" (Yepsen, 1945, p. 96).

In 1942, *Science* published Goddard's article, "In Defense of the Kallikak Study." In the article, he responded to Meyerson, Scheinfeld, and other critics. He commented that for more than a decade the study was accepted without question but that as time went by critics arose who, he charged, "obviously had not read the originals, and who therefore thought that they detected certain flaws in the techniques which did not exist" (Goddard, 1942, p. 574). To Meyerson's skepticism of the ability of field workers to diagnose feeble-mindedness accurately, he responded:

Not understanding the purpose or the methods of the field-worker, Dr. Meyerson makes his own assumptions. He argues that because he cannot correctly diagnose feeble-mindedness, nobody can. Therefore all our diagnoses must be guesses and "surmises."

The record shows that our field-workers were carefully trained. . . . They spent weeks and months in the institution, talking with and observing all grades of defectives. It is well known that superintendents of such institutions quickly learn, and when a new arrival appears they not only know whether he is a fit subject for their institution or is normal and does not belong there, but they also know his grade. Even the attendants acquire this ability rather quickly. Dr. Fernald used to enjoy telling how his attendants would spot a child on the train and report that a new case was on the way." (pp. 574–575)

In the article, Goddard then turned his attention to Meyerson's comments concerning the mother of Martin, Jr. He said that Meyerson

ridicules the idea that we could know that the mother of the Kallikaks was feeble-minded, when we "did not even know her name," but had to put her down as "nameless." I did not realize that it might mislead. All names are fictitious, and it occurred to me that "nameless" would identify her without any possibility of confusion. She is nameless to the reader only. We had her name; and not only her name but her history. We were fortunate enough to find an intelligent lady of advanced age, who knew personally the "nameless one." That seems impossible until one realizes that if each of them had lived to be 80, they could have known each other for eight years or more; and if they had lived to be 90—not impossible—they could have been neighbors for 30 years! (p. 575)

In fact, we know from Goddard's correspondence with Elizabeth Kite that he did not know the name of the girl. He asked Kite if she knew the name and explained that he wanted it for just this sort of "table turning" on his critics. As noted in the chapter on Elizabeth Kite, she was unable to provide a name. In 1913, she admitted that information about the life of Martin, Jr.'s, mother and her level of intelligence was based on little direct evidence. Her feeblemindedness was assumed from the lives of her descendants and the vague second-hand recollections of a few people who had never met her. Kite said, "I can get no one who remembers her, though I found several people who remember that their mothers recognized something about her different from other women and they talked about her a great deal" (Kite, 1913, pp. 151–152). This is hardly a convincing basis for a sound diagnosis, and it clearly contradicts Goddard's assertions.

In response to Scheinfeld's statement that the study rested largely on the assumption that Martin, Sr., was the father of the illegitimate child and that no court would accept as evidence of paternity the information supplied in the book, Goddard's brief reply was: "A strange statement. Courts have always accepted such evidence and still do. In this case there was not even a doubt." (Goddard, 1942, p. 575)

Concerning Scheinfeld's question of the genetic mechanism by which Martin, Sr., could have sired a defective son, Goddard wrote:

Certainly Martin Kallikak, Sr. must have been a "Simplex," else his son by the "Nameless" would have been normal. But that is no argument. It is well known that a trait may remain recessive for generations as long as its possessors mate with "duplexes." (p. 575)

Goddard closed his article by charging that his critics did not have an adequate understanding of what they were criticizing. He implored them to read and analyze carefully the original work for their own protection and for the "preservation of truth and the advancement of science" (p. 576). (A photograph of Goddard at about this time is shown in Illustration 8.)

Goddard apparently sent copies of the *Science* article to a large number of his former students and colleagues. A file in the archives at the University of Akron contains more than 20 letters of thanks for the article and support for his defense. Letters from Lewis Terman and Edgar Doll are included. From Doll:

Congratulations on your excellent statement "In Defense of the Kallikak Study." Thank you for sending me the copy of *Science* in which it appears. You covered the ground effectively, temperately, and without rancor. I think the statement was timely as well as necessary, and rather more effective than it might have been at some earlier date.

You might like to know that this was called to my attention by several of the psychologists who were here two weeks ago and that their comment was definitely favorable to your position and fairness of statement. Knowing your reluctance to indulge in controversial discussion I can appreciate the hesitation with which you may have prepared this. (Doll, 1942, p. 124)

As Goddard's successor at Vineland's research laboratory, Doll had followed Goddard's conceptual tradition of viewing mental retardation as primarily genetic in origin. In his correspondence, he states:

Briefly my personal view is that heredity is the most important single factor in the causation of mental deficiency. My general opinion is that idiocy, low-grade imbecility, and the majority of clinical types of mental

Illustration 8: Henry Goddard in 1941. Reprinted with permission from the archives at Ohio State University.

deficiency are not hereditary, that is, not inherent in germ plasm. However, numerically these degrees and types do not constitute a large portion of all the feeble-minded. The vast majority of the feeble-minded are in the moron grade, and there is very little information outside the field of heredity to explain the cause of their condition. (Doll, 1937, p. 137)

In 1944, Scheinfeld countered Goddard's defense of the Kallikak study. He reacted to the points Goddard had raised and made a strongly worded assessment of the social impact of the work. Here are excerpts from his article in the *Journal of Heredity:*

But the ghosts of the notorious Kallikaks have not easily been exorcised . . . as synonymous with all that is degenerate in the human germ plasm, they go marching on through many textbooks and reference works (some published in the last few years) and tens of thousands of college and high school students continue each year to be taught the lessons of these horrible examples. . . .

But today the Kallikak study has acquired new implications. For the premise set forth—that there are genetically "superior" and "inferior" categories of human beings, clearly defined as such through unitary traits based on single gene differences—is one which underlies group concepts which have helped to bring on the present war (as they threaten to bring on other wars), and which have created many bitter conflicts within our own and other countries. We therefore are less interested in the Kallikak study itself than in the way it has been or may be applied to broader fields of thought; and it thus becomes more than ordinarily important at this time to set the record straight regarding the scientific validity of Dr. Goddard's findings and conclusions. . . .

Indeed, any complicated study of this kind, made in the infancy of genetics and with the rudimentary psychological techniques of thirty years ago, would be expected to have many flaws, and under ordinary circumstances it would be unfair to expose it to the glaring light of present-day analysis. Unfortunately, Dr. Goddard has challenged such inspection by his implication that time has not withered nor new knowledge staled his procedures and findings. . . .

Dr. Goddard credits this writer with having contributed "one original idea": that proof of the Kallikak dichotomy, to quote from *You and Heredity*, "rests largely on the assumption that the illegitimate child whom the feebleminded mother chose to call 'Martin Kallikak, Jr.' was indeed the son of the man she designated, which no court would accept as evidence. . . ."

We continue to doubt that modern jurists would accept as valid proof in a hypothetical legal case today the uncorroborated claim of a reputedly feebleminded tavern girl of 150 years ago as to the paternity of her illegitimate child. In any event, it still remains for geneticists to decide whether such evidence would be acceptable as the point of departure for a valid "scientific experiment." If not, then the whole dichotomous aspect of the Kallikak study, balanced precariously like a huge, inverted pyramid with its apex on this single point of the dual mating, topples of its own weight. . . .

Dr. Abraham Meyerson, in his analysis of the Kallikak case some twenty years ago, intimated that the diagnoses may have rested largely on "surmises." Nevertheless, a careful reading of the Kallikak study must arouse suspicion that once the premise of two distinct kinds of Kallikaks had been established, the investigator had set out with two different intellectual paintpots, to gild the *Lilium candidum* and to tarbrush the *Spathyema foetidus*. . . .

However one might regard this technique of appraising the intelligence of long-dead individuals, the facility with which Dr. Goddard or his investigator could make their diagnoses may explain another remarkable fact, revealed to many of us for the first time through his recent communication: That the entire Kallikak study . . . [was] begun and completed between 1910 and the publication date in September 1912. Contemplating the presumably colossal job of ferreting out the case histories and establishing the mental grades of a thousand or more individuals for five generations back, present-day research workers must either marvel at the speed with which this was accomplished, or else must question the thoroughness with which the study was made. . . .

What shall be said of Dr. Goddard's assumptions or conclusions that all such conditions in the bad Kallikaks as sexual immorality, alcoholism, pauperism, epilepsy, and criminality, as well as mental defect, were all related manifestations of the same genetic weakness, a unitary condition determined by or resulting from the same recessive gene which produced feeblemindedness? Or of his opinion, in analyzing the fact that the first batch of "bad" Kallikaks had resulted from a mating of the illegitimate Martin, Jr. with a *normal* woman, that this must be considered as "demonstrating that the defect is transmitted through simple recessive genes, then the progenitor of the dichotomy, the soldier who later married the Quakeress, must have been heterozygous for these genes." So by what good fortune was not this bad gene passed along to any of the numerous "good" Kallikaks, or ever manifested itself in defect among any of them? Surely the laws of chance must have awarded some of the seven good Kallikaks the shady half of their father's "demonstrably" mixed heredity. . . .

Where criticism does seem to be justified is in his failure to consider the possibility that differences in environment might have been strong factors in creating at least some of the disparity between the two Kallikak branches. . . .

What should interest us now is why, in view of the easily apparent flaws in the Kallikak study, and its rejection in authoritative circles for many years, it has continued to be given such strong credence and to find such warm support in many quarters? The answer is a simple one. As suggested previously, there are persons everywhere who relish the thought that some groups, races, classes or strains (always including the ones to which they themselves belong) are born to be superior and dominant, and that other groups are destined by nature to be inferior and subordinate. Thus, the Kallikak study when it appeared was eagerly welcomed because it apparently offered "scientific proof" that a high proportion of the social and physical ills of mankind were directly or indirectly due to hereditary defects, and that these could be eliminated most effectively and a super race speedily produced, by breeding out the "unfit." No one in possession of the facts can doubt the existence of pathological genes in human germ plasm. It is the unjustified extension of pseudogenetic principles into sociology that is a danger. Social action based on such unsound premises can be very dangerous.

Certainly, Dr. Goddard cannot be held responsible for the misuses of his study, nor should this article be construed as in any sense directed against him personally. . . .

The fact is that after more than three additional decades of research by innumerable investigators, there is much more uncertainty as to the diagnosis, etiology and genetic aspects of the various types of mental and social deficiency than was evidenced in the sweeping generalizations and pat conclusions of Dr. Goddard's reports. . . .

Perhaps some other investigators, equipped with more modern techniques and approaches, will find it of interest to take up where he left off, to dig further into these pedigrees and produce more scientifically valid proof. But until this happens we will have to nurse the suspicion that if all the bad little Kallikaks had been brought up in exactly the same environment as was accorded to all the good little Kallikaks, the distinctions between the two groups might not have been so glaringly marked, and not nearly so many of the bad Kallikaks would have toppled from their places, or fallen by the wayside. Nor, might the Kallikak study itself have fallen down so sorely had it given the bad Kallikaks a fairer break. (Scheinfeld, 1944, pp. 259–264)

The most recent critique of the Kallikak study is in Stephen Jay Gould's brilliant book, *The Mismeasure of Man*. After describing the study and delineating its weaknesses, Gould makes a startling revelation:

It may be a small item in the midst of such absurdity, but I discovered a bit of more conscious skulduggery two years ago. My colleague Steven Selden and I were examining his copy of Goddard's volume of the Kallikaks. The frontispiece shows a member of the kakos line, saved from depravity by confinement in Goddard's institution at Vineland. Deborah, as Goddard calls her, is a beautiful woman. She sits calmly in a white dress, reading a book, a cat lying comfortably on her lap. Three other plates show members of the kakos line, living in poverty in their rural shacks. All have a depraved look about them. Their mouths are sinister in appearance; their eyes are darkened slits. But Goddard's books are nearly seventy years old, and the ink has faded. It is now clear that all the photos of non-institutionalized kakos were phonied by inserting heavy dark lines to give eyes and mouths their diabolical appearance. The three plates of Deborah are unaltered.

Selden took his book to Mr. James H. Wallace, Jr., director of Photographic Services at the Smithsonian Institution. Mr. Wallace reports (letter to Selden, 17 March 1980):

There can be no doubt that the photographs of the Kallikak family members have been retouched. Further, it appears that this retouching was limited to the facial features of the individuals involved— specifically eyes, eyebrows, mouths, nose and hair.
By contemporary standards, this retouching is extremely crude and obvious. It should be remembered, however, that at the time of the original publication of the book, our society was far less visually sophisticated. The widespread use of photographs was limited, and casual viewers of the time would not have nearly the comparative ability possessed by even preteenage children today. . . .
The harshness clearly gives the appearance of dark, staring features, sometimes evilness, and sometimes mental retardation. It would be difficult to understand why any of this retouching was done were it not to give the viewer a false impression of the characteristics of those depicted. I believe the fact that no other areas of the photographs, or the individuals have been retouched is significant in this regard also. . . .
I find these photographs to be an extremely interesting variety of photographic manipulation. (Gould, 1981, p. 171)

NOTES

1. It is likely that Goddard had suggested the use of feebleminded actors for certain parts in the play. Goddard felt that the feebleminded were excellent mimics and had been a strong supporter of dramatics while at the Vineland Training School.
2. In 1914, Dr. John H. Kellogg, creator of the breakfast cereals for which his family became famous, established and endowed the Race Betterment Foundation. Through the influence of Charles B. Davenport, Kellogg had become interested in promoting the cause of eugenics. The Race Betterment Foundation was headquartered in Battle Creek, Michigan, and held frequent conferences there (Voorhees, 1981).

REFERENCES

(1913). *American Journal of Psychology, 24,* 290–291.

(1912, October 1). *Dial,* p. 247.

Doll, E. (1937). [Letter]. *Doll papers* (Box M236, Correspondence, p. 137). Akron, Ohio: University of Akron, Bierce Library, Archives of the History of American Psychology.

Doll, E. (1942). [Letter]. *Doll papers* (Box M233, Correspondence, p. 124). Akron, Ohio: University of Akron, Bierce Library, Archives of the History of American Psychology.

Dunlap, K. (1940). Antidotes for superstitions concerning human heredity. *Scientific Monthly, 51,* 221–225.

East, E.M. (1927). *Heredity and human affairs.* New York: Scribner.

Goddard, H.H. (1913a). [Letter]. *Goddard papers* (Box M615, Publishers). Akron, Ohio: University of Akron, Bierce Library, Archives of the History of American Psychology.

Goddard, H.H. (1913b). [Letter]. *Goddard papers* (Box M615, Publishers). Akron, Ohio: University of Akron, Bierce Library, Archives of the History of American Psychology.

Goddard, H.H. (1929). [Letter]. *Goddard papers* (Box 35.1, Correspondence). Akron, Ohio: University of Akron, Bierce Library, Archives of the History of American Psychology.

Goddard, H.H. (1942). In defense of the Kallikak study. *Science, 95,* 574–576.

Gould, S.J. (1981). *The mismeasure of man.* New York: W.W. Norton.

(1912). *Independent, 73,* 794.

James, D.L. (1925). [Letter]. *Goddard papers* (Box M33.1, AA4(1)). Akron, Ohio: University of Akron, Bierce Library, Archives of the History of American Psychology.

James, D.L. (1926). [Letter]. *Goddard papers* (Box M33.1, AA4 (1)). Akron, Ohio: University of Akron, Bierce Library, Archives of the History of American Psychology.

Kauser, A. (1913). [Letter]. *Goddard papers* (Box M615, Publishers, p. 121). Akron, Ohio: University of Akron, Bierce Library, Archives of the History of American Psychology.

Kite, E.S. (1913). Mental defect as found by the fieldworker. *Journal of Psycho-Asthenics, 18,* 145–154.

McCallie, J.M. (1941). [Letter]. *Goddard papers* (Box M32, Ephemeris, p. 5). Akron, Ohio: University of Akron, Bierce Library, Archives of the History of American Psychology.

Meyerson, A. (1925). *The inheritance of mental disease.* New York: Williams & Wilkins.

(1913). *Popular Science Monthly, 82,* 415–416.

Scheinfeld, A. (1939). *You and heredity.* New York: Stakes-Lippincott.

Scheinfeld, A. (1944). The Kallikaks after thirty years. *Journal of Heredity, 35,* 259–264.

Van Wagenen, B. (1913a). [Letter]. *Goddard papers* (Box M615, Correspondence). Akron, Ohio: University of Akron, Bierce Library, Archives of the History of American Psychology.

Van Wagenen, B. (1913b). [Letter]. *Goddard papers* (Box M615, Correspondence). Akron, Ohio: University of Akron, Bierce Library, Archives of the History of American Psychology.

Voorhees, D.W. (1981). (Ed.). *Dictionary of American biography.* New York: Charles Scribner's Sons.

Yepsen, L.N. (1944). [Letter]. *Goddard papers* (Box M42, G7). Akron, Ohio: University of Akron, Bierce Library, Archives of the History of American Psychology.

Yepsen, L.N. (1945). [Letter]. *Goddard papers* (Box M42, G7). Akron, Ohio: University of Akron, Bierce Library, Archives of the History of American Psychology.

Chapter 7

Revisiting the Kallikaks

As noted earlier, I have been aware of the Kallikak study since I was a college student majoring in psychology. The study was first presented to me in a blend of humor and criticism, and I have discussed the book in a similar fashion with my own students. While vaguely aware all the while of its impact on the treatment of mentally retarded people, I viewed the Kallikak work primarily as an example of primitive and naive psychological research. It was not until I read Stephen Gould's *The Mismeasure of Man* that I was inspired to look more seriously at the Kallikak study and the circumstances surrounding it. Gould's description of Goddard's work—in particular, his disclosure that photographs in the

Kallikak book had been altered to produce a more sinister look to the "bad" side of the family—reawakened in me images of Deborah and her ancestors. The more I thought of them, the more I felt compelled to try to discover all that I could about them. If something could be found that would bring into question the actual substance of the story of the bad seed, not just the conclusions drawn from it, it might be a valuable scholarly contribution. More importantly, it might ease my mind.

How do you go about finding the characters in a story that was written 75 years ago? To add to the difficulty, the family name you have to work with is a pseudonym. Given names and place names have also been changed. The only certainties are that the family lived somewhere in New Jersey and that Deborah was at the Vineland Training School.

I first went to the Kallikak book itself. From Goddard's cautious description of members on the "good" side of the family and of the successful families into which they had married, I tried to find these members in the genealogies of families prominent in New Jersey history. Goddard had described in glowing terms the descendants of Martin Kallikak, Sr., and his wife who had married into families descended from "signers of the Declaration of Independence" and "the founder of a great university." I felt that perhaps with these and other clues from the book I could find the family name that would allow me to break the code. This approach at first generated the excitement of being on the right track, of getting close; but finally it led to the disappointing realization that I would never find the family through that sort of best-guess method. Yet, I was in fact close. I had come to believe that the Stockton family was somehow related, and for a while I thought that the Kallikaks were Stocktons. As it turns out, the Kallikaks were related through marriage to descendants of Richard Stockton (a signer of the Declaration of Independence) and through them to Benjamin Rush (another signer). The Stockton family was also instrumental in the founding of Princeton University.

I decided to attempt to contact people who had been acquainted with Goddard and who, thereby, might know the family name. I wrote to several people who were former students, colleagues from Goddard's later years at Ohio State University, or members of families that had had some connection with the Vineland research. From most of them I received no reply. I mention this because the failure to respond in this case proved to be characteristic of the reluctance of those who knew Goddard to talk about his work. I cannot, of course, interpret this reluctance, but I suspect that it may have been based on feelings about the vulnerability of his work and reflected an attempt to protect Goddard from criticism.

One person who did reply is a well-known and respected psychologist who knew Goddard late in the latter's career. She was encouraging when I explained my reasons for wishing to revisit the Kallikak family. She provided me with helpful information and insights concerning Goddard's life and work. However, she stopped short of giving me the name of the family, which she apparently knew, possibly because of her lack of certainty about my intentions. She also seemed somewhat concerned about Goddard's image. I feel that I must withhold her name at this point out of respect for her, but I appreciate the help that she gave me in my quest for the Kallikaks and I wish to express that appreciation here.

I next turned to census figures, archival collections, and historical documents in search of patterns that would fit Goddard's story. I had the feeling that, if I searched diligently enough, the Kallikaks would suddenly emerge from some dusty volume. In the end, however, my discovery of the real name of the Kallikak family did not come from my searches in libraries, archives, or historical societies. I eventually would return to these sources for the real flesh of the Kallikak story, but initially they were not helpful to me. Without some name as a starting point, there was little I could find in them that would assist me in identifying the family. It was a double bind: Without a family name I had little chance of finding the family through such sources.

I decided that, as a starting point, I should focus my attention on one member of the family. The logical choice was Deborah. If I could identify her, it would open up the whole family to me. Unfortunately, most of the information on Deborah was restricted. Material concerning her that was inventoried in Goddard's papers or the Vineland research files was classified as medical information on a former patient of a state facility and was therefore confidential. At several points I was fairly certain that documents that would have revealed Deborah's identity to me were in a file only a room away yet I was not allowed to see them. It was disconcerting, but I must express my admiration to the archivists and librarians whose ethics prevailed in the face of my frustration.

Deborah's identity was finally revealed to me through a quite different source, through friendship. Quite discouraged from my archival searches, I decided to visit Vineland. Knowing that any records there would also be considered confidential, I did not appeal to the training school (now the American Institute for Mental Studies) or the Vineland State School to open their files on Deborah to me (later I did ask for assistance from the state school but received no reply). Instead I tried to get a feel for the atmosphere of the town and what it must have been like during Deborah's years there—an attempt, I suppose, to feel the "spirit of the place." I was hoping too that I might just stumble onto something that would help. It was during my ramblings in Vineland that—through accident, good fortune, or both—I met a man and woman who had known Deborah. Serendipity! They were not mental retardation professionals, archivists, librarians, or historians—but they gave me Deborah. They were ordinary citizens of Vineland who, through their voluntary concern and efforts, had known Deborah until her death in 1978. I met them through chance, our conversations led to Deborah, and they delighted in telling me about her. And they gave me her name. My search now had direction.

With the family name, I began research again with census figures, legal records, genealogical information, and historical documents. Work at the National Archives, at the Library of Congress, and in courthouses in New Jersey resulted in a steady unveiling of the Kallikaks: They were really there! More importantly, they were there with a difference. The Kallikaks that I found differed greatly from the Kallikaks Goddard had described. They were the same individuals, but what the records told me about them was often contrary to what Goddard's book had portrayed.

Once I discovered Deborah's real name, tracing the rest of the family became a matter of persistence in finding relevant sources of information and knowledgeable people who could help me. I was very lucky in both respects. The people whom I have acknowledged at the beginning of the book are those who made possible my most interesting discoveries.

I gave much thought to whether or not I should reveal the true family name of the Kallikaks. I came to the conclusion that it would serve no reasonable purpose. I do not want this book to rest on the exposure of a name. In the following discussions, therefore, I continue to use the name Kallikak, even for descendants whose names are different through marriage. I feel that this will help readers avoid confusion and make it easier to trace the supposed "bad seed" through the generations. Because I feel it is necessary to document some of what I present here, it will now be much easier for other interested parties to discover the real name if they wish to do so. By continuing to use the pseudonym in this book, however, I believe I can ensure that finding the name will at least require a serious effort. Goddard said that he protected the name for the sake of the "good" family; he felt that the "bad" family would be oblivious to its use anyway. I am cautious with the name for the sake of those who I choose to think were "disfavored" by the story, the so called "bad" Kallikaks. Although I think that what I have found vindicates them, many of the descendants of that side of the family are surely unaware of their family's portrayal in the study and should not be needlessly troubled by it. The "favored" Kallikaks will not be damaged by anything presented here. Should they become aware of their role in the Kallikak story, the worst that would be required of them would be a slight alteration in their family tree—perhaps not even that if they choose to ignore a pruning that occurred many years ago.

The findings presented in this chapter are documented differently from those in previous chapters. Here I limit the documentation in cases where a full reference would directly or indirectly disclose the family name. I describe sources in the narrative but stop short of giving page numbers or other information that would make the family name easily accessible. At some points, for the same reason, I alter the names of people and places. I walked a mental tightrope on this issue for many nights and decided to make this compromise. I believe it allows me to be responsible in my scholarship and at the same time refrain from making public the Kallikak name. At least that is my intent.

The obvious starting point for revisiting the Kallikaks is Martin, Sr. According to Goddard,

> When Martin Sr., of the good family, was a boy of fifteen, his father died, leaving him without parental care or oversight. Just before attaining his majority, the young man joined one of the numerous military companies that were formed to protect the country at the beginning of the Revolution. At one of the taverns frequented by the militia he met a feeble-minded girl by whom he became the father of a feeble-minded son. . . . Martin Sr., on leaving the Revolutionary Army, straightened up and married a respectable girl of good family, and through that union has come another line of descendants of radically different character. (Goddard, 1912, p. 18)

Later in the book, Goddard goes into more detail, explaining that Martin, Sr.

> joined the Revolutionary Army in April, 1776. Two years later he was wounded in a way to disable him for further service, and he then returned to the home farm. During the summer of enforced idleness he wooed and won the heart of a young woman of good Quaker family. Her shrewd old father, however, refused to give his consent. To his objections, based on the ground that Martin did not own enough of this world's goods, the young man is recorded as saying, "Never mind. I will own more land than ever thou did, before I die," which promise he made true. That the parental objection was overruled is proved by the registry of marriage, which gives the date of Martin's union with the Quakeress as January, 1779. (Goddard, 1912, pp. 29, 99–100)

Goddard's description of Martin, Sr., gives the impression of a young soldier far from home who becomes involved with a defective tavern girl out of loneliness. After his service he returns home, repents of his loose morals, and marries a girl of equal station in life. They raise an upstanding family together. This portrayal is essentially correct, but a few details change the tone of the circumstances.

According to the Revolutionary War pension records housed in the National Archives, Martin, Sr., did in fact enlist in April of 1776. He joined the New Jersey militia as a private in the Second Hunterdon County Regiment under a Captain Ely. His service was, however, neither long nor far from home. He served monthly tours, every other month, over a period of a year and a half. He was back living at home every second month. Even during his months of service he was never far away. His duty consisted of standing guard in Hunterdon County. Martin, Sr., never saw combat during the Revolution. The records indicate that he was

injured in the right arm by the accidental discharge of a
musket, shortly before the Battle of Monmouth.

Martin, Sr.'s, lawful wife was named Rachel. There is
some confusion as to when he and Rachel were married.
The *Marriage Record of the Ancient Parish* gives his marriage date as January, 1779. The Revolutionary War pension
records indicate that he was married in July of 1776. Martin,
Jr., was born in 1776. Martin, Sr.'s, first legitimate child,
Frederick, was born on April 22, 1779. Either Frederick was
born very early (three months) in the marriage or the illegitimate Martin, Jr., was born shortly before or shortly after his
father's marriage to Rachel.

In 1833, Martin, Sr., applied for and was granted a pension for his military service. He died in 1837. Rachel continued to receive the pension until her death in 1842.

Martin, Sr., accumulated a great deal of real estate during
his life, and his holdings were passed down through the
family. His legitimate children were indeed favored by the
environment provided by their parents. Through the generations, they have tended to be financially successful, well-
educated, and socially prominent. A genealogy of the
favored Kallikaks, published in 1932 by a great granddaughter of Martin, Sr., portrays them as a productive and
respected line. The material included in that publication is
alluded to by Goddard in the Kallikak study. Of its author, he
writes:

> This lady is a person not only of refinement and culture
> but is the author of two scholarly genealogical works.
> She has, for years, been collecting material for a similar
> study of the Kallikak family. This material she gener-
> ously submitted to the use of the field worker. In the end
> she spent an entire day in the completion and revision of
> the normal chart presented in this book. No praise can
> be too high for such disinterested self-forgetfulness in
> the face of an urgent public need. (Goddard, 1912,
> pp. 98–99)

For those who have wondered over the years how Goddard collected information on the hundreds of people
described in the Kallikak book in such a short time, here is at
least part of the answer. The data that he presents on the
favored side of the family was taken directly from the manuscript of the great granddaughter of Martin, Sr. Even the
wording in certain sections of Goddard's book is identical to
that used in the earlier manuscript. It is inconceivable that in
the great granddaughter's genealogical research on the
favored Kallikaks she would have not come across and
recorded some data on the disfavored line. There is a good
chance, I think, that she also provided Goddard and Elizabeth Kite with some of the information on the descendants
of Martin, Jr. The resulting line of descent from Martin, Jr., to
Deborah is shown in Illustration 9.

Goddard's description of Martin, Jr., is laden with those
traits he felt characterized the moron:

> In 1803, Martin Kallikak Jr., otherwise known as the "Old
> Horror," married Rhoda Zabeth, a normal woman.
> They had ten children, of whom one died in infancy and
> another died at birth with the mother. . . . He was
> always unwashed and drunk. At election time, he
> never failed to appear in somebody's cast-off clothing,
> ready to vote, for the price of a drink, the donor's ticket.
> . . . Simple . . . not quite right, but inoffensive and kind.
> All the family was that. . . . Old Martin could never stop
> as long as he had a drop. Many's the time he rolled off of
> Billy Parson's porch. Billy always had a barrel of cider
> handy. He'd just chuckle to see old Martin drink and
> drink until finally he'd lose his balance and over he'd
> go! (Goddard, 1912, pp. 19, 51, 61, 80)

According to census data for Hunterdon County, Martin,
Jr., was born in 1776. I was unable to find any information
concerning his childhood. There is nothing to indicate that
he was ever acknowledged or supported in any way by his

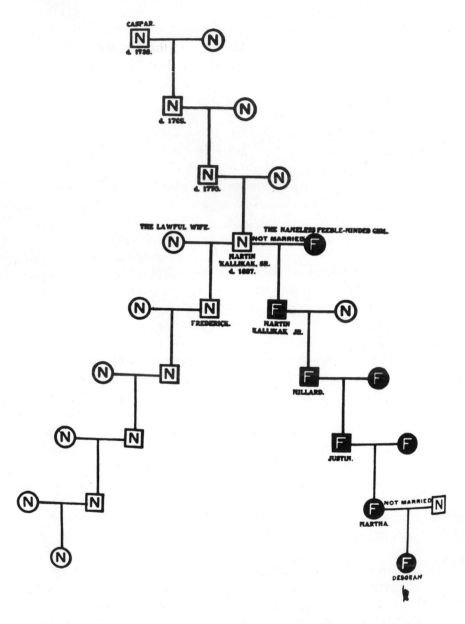

Illustration 9: Primary line of descent from Martin, Jr., to Deborah. Reprinted with permission from *The Kallikak Family: A Study in the Heredity of Feeble-Minded-ness* (p. 36) by H.H. Goddard, 1912, New York: Macmillan.

father, even though he carried his father's unusual name and both lived their entire lifetimes in the same county.

Martin, Jr., married the woman that Goddard called Rhoda Zabeth in October, 1804. They remained together for the next 22 years, until her death. The *Hunterdon County Gazette* reported in November of 1826: "Died in this Township on the night of the 8th instant, after a severe illness of a few days, Rhoda Kallikak, wife of Martin Kallikak."

Martin, Jr., owned land throughout most of his adult life. County records indicate that he purchased two lots of land in 1809 for cash. Deed books for the county contain records of his transfer of his property to his children and grandchildren later in his life. The 1850 census record shows that he was living with one of his daughters and several of his grandchildren at that time. That record also lists all of the adults in the household as being able to read. The 1860 census record lists his occupation as "laborer" and his property as valued at $100 (not a meager amount for the average person at that time). Martin, Jr., died in 1861.

Goddard devoted considerable attention to three of Martin, Jr.'s, daughters as examples of the inevitable degeneracy of the moron's bad seed.

> Martin Jr.'s fourth child, "Old Sal" was feeble-minded and she married a feeble-minded man. Four of their children are undetermined, but one of these had at least one feeble-minded grandchild. . . . The two other children of Old Sal were feeble-minded, married feeble-minded wives, and had large families of defective children and grandchildren. (Goddard, 1912, pp. 21, 79)

Thus, Sal Kallikak is presented as a moron and the mother of morons. (Retouched photographs from the Kallikak book showing the grandchildren of "Old Sal" are displayed in Illustration 10.) However, a family history of Sal's descendants reveals many contradictions to this portrayal. Two of her grandchildren are still living. A brother and sister, they

Illustration 10: Retouched photographs that give a sinister look to the grand-children of "Old Sal." Reprinted from *The Kallikak Family: A Study in the Heredity of Feeble-Mindedness* by H.H. Goddard, 1912, New York: Macmillan.

94

are both retired school teachers living in Trenton, New Jersey. One grandson moved from New Jersey to Iowa, became treasurer of a bank, owned a lumber yard, and operated a creamery. Another grandson moved to Wisconsin. His son served as a pilot in the Army Air Corps in World War II. A great, great grandson of Sal Kallikak is a teacher in Chicago. A great grandson was a policeman in another city in Illinois.

A 1930 newspaper article reports that all of Sal's sons were soldiers in the Civil War. The article was written by a man who had known Sal's sons while they were in school. In writing of a boy who had been in school with one of the sons, he notes that "he and Aaron Kallikak were great cronies, often associated in harmless escapades." He describes Sal's son in these words:

> Aaron Kallikak was a man of good mind and something of a student in his way. At one time he was very much interested in physiognomy, which was then a local fad among students. He reached what he thought sufficient skill for a venture into the lecture field. So he posted notices of "A Lecture on Physiognomy by Prof. A.H. Kallikak," to be held in Hardscrablle School House. Several of us went out to hear what the "professor" had to say. . . . Aaron looked the crowd over, seemed to consider for a short time, then rose and said in his stammering way: "The lecture will not be given tonight. I want an appreciative audience."
>
> Then we knew exactly what the "professor" had to say under the circumstances. We came away with increased respect for his self-suppressed ability and his keen thrust in what he thought was self defense. (*Hunterdon Democrat Advertiser*)

Goddard's profiles of the other daughters of Martin, Jr., are just as negative as that of Sal:

Illustration 11: Retouched photograph of Malinda, the daughter of Jemima. Reprinted with permission from *The Kallikak Family: A Study in the Heredity of Feeble-Mindedness* by H.H. Goddard, 1912, New York: Macmillan.

The fifth child of Martin Jr. was Jemima, feeble-minded and sexually immoral. She lived with a feeble-minded man named Horser, to whom she was supposed to have been married. Of her five children, three are known to have been feeble-minded, two are undetermined. (Goddard, 1912, p. 21.)

A retouched photograph from the Kallikak book showing one of Jemima's daughters is presented in Illustration 11.

Goddard's profile of another daughter is in the same vein:

The sixth child of Martin Jr., . . . known as "Old Moll" was feeble-minded, alcoholic, and sexually immoral. She had three illegitimate children who were sent to the almshouse, and from there bound out to neighboring farmers. . . . Old Moll, simple as she was, would do anything for a neighbor. She finally died—burned to death in the chimney corner. She had come in drunk and sat down there. Whether she fell over in a fit or her clothes caught fire, nobody knows. She was burned to a crisp when they found her. (Goddard, 1912, pp. 21, 79)

Jemima was born in Hunterdon County and was still there when she died at age 86 in December, 1900. She was indeed married (not "supposedly"); according to a newspaper report, she was the widow of a well digger in the vicinity. In 1900, she had come into the town of Flemington to visit her daughter when she became ill; she was thought to be "on the mend" when she died (*Hunterdon Democrat Advertiser*, 1900). The 1860 census record shows her living at that time with her husband John, who was 50. I was unable to find information on her children.

"Old Moll" was also married and was living with her husband during the 1850 census. The story of her being burned to death is true and was reported in the *Hunterdon Democrat* in 1853.

A further clue to how Goddard and Elizabeth Kite collected their information and how they arrived at charac-
terizations like those of Martin, Jr.'s, daughters is contained
in the correspondence of Hiram Deats, founder of the Hunterdon County Historical Society. Russell Bruce Rankin, editor of *The Genealogical Magazine of New Jersey*, wrote to
Deats on March 5, 1941:

> I once read the Kallikak Family, but never found anyone
> who knows what family it is supposed to represent.
> Neither could I figure out how anyone could write such a
> complete genealogy of such a peculiar outfit, particu-
> larly on the dark side of the picture.

In his reply of March 8, 1941, Hiram Deats included the
following comments:

> It is doubtful if anyone ever identified the real name of
> the original of the Kallikak family. The descendants had
> a lot of names, and none of them that I knew was the
> original name. The genealogy of the good part of the
> family was published and you have seen it. Dr. God-
> dard insisted on protecting them, even after a century
> and a quarter, and more, so out of respect to him, I have
> never mentioned it to anyone. I was District Clerk at the
> time, of Raritan township, and we had two of the family
> in one school, and one in another. I had appointed
> myself truant officer, and tried to get some regular atten-
> dance out of them, though they were hardly fit for the
> Feeble Minded Institution at Vineland. Then when Dr.
> Goddard's assistant came, wanting help, I felt it might
> result in getting them out of the township, so gladly
> helped. Spent a lot of time going over court records and
> Justice dockets. But I do not care to do such a job again. I
> can still see those eyes. "Eyes have they, but they see
> not," etc.

Deats was a gentleman farmer who left the operation of his farm to hired workers while he devoted his time primarily to genealogy and local history. He was obviously more interested in the possibility of ridding the county of what he considered "undesirables" than in finding the truth. I have found no records of what he actually reported concerning the Kallikaks or how extensive his contributions may have been.

The line of descent from Martin, Jr., to Deborah was through the son that Goddard called Millard:

Millard [was] the direct ancestor of our Deborah. He married Althea Haight and they had fifteen children. . . . Millard married Althea Haight about 1830 . . . the mother died in 1857. . . . This mother, Althea Haight, was feeble-minded. That she came from a feeble-minded family is evidenced by the fact that she had at least one feeble-minded brother, while of her mother it was said that the "devil himself could not live with her. . . ." Millard Kallikak married for his second wife a normal woman, a sister of a man of prominence. She was, however, of marked peculiarity. (Goddard, 1912, pp. 19, 23, 26)

Millard was a cooper (barrelmaker). He died in 1893 at 90 years of age. He was Martin, Jr.'s, oldest son and the father of Justin, Deborah's grandfather. Millard owned a 46-acre tract of land in Hillsborough Township, Somerset County. He and Althea had, as indicated, 15 children.

The 1850 census record shows him owning real estate valued at $250. By 1860, the census record lists him as owning real estate valued at $1,600.

Millard's second wife was Mary. She was still alive in June 1900, and was living with her sister, Margaret, in Montgomery Township, Somerset County. Millard and Mary had three children.

The *Unionist-Gazette* of Somerville, New Jersey, reported Millard Kallikak's death in June of 1893: He "died on Wednesday last and was buried on Saturday. He was a good Christian and well thought of by all who knew him." Hardly the epitaph of a moron.

The grandfather of Deborah was Justin Kallikak. He is briefly described by Goddard: "The third child of Millard was Justin, the grandfather of our Deborah. . . . He was feeble-minded, alcoholic, and sexually immoral. He married Eunice Barrah, who belonged to a family of dull mentality" (Goddard, 1912, pp. 24, 26).

With Justin, the lives of the Kallikaks took a dramatic turn. He and his family moved to an area just outside Trenton. Up to that point, Martin, Jr., and his descendants had lived in the rural and small-town atmosphere of Hunterdon and Somerset counties. Though many of them had lived with limited resources and against considerable environmental odds, the records suggest that they were a cohesive family. The change to a more urban environment—and also perhaps the changing times—apparently imposed a stress on this tradition of cohesion and support. The existing records suggest that Justin was unable to hold his family together. Following the death of his wife, Justin's children were taken into other families or were left to manage on their own.

Justin Kallikak was born in 1832; he married Eunice Barrah when he was 30 years old. Eunice was 8 years younger than Justin. The 1870 census records show them living in the rural environment where the family had lived for generations. The entry indicates that they had four children living at home with them. By 1880, Justin and Eunice had moved to the growing urban and industrial sprawl of Trenton. The census records for that year show that none of their children were living with them. Justin's occupation is given as "laborer." Eunice and Justin eventually had 11 children, 6 of whom died in infancy. After Eunice's death, Justin married a woman that a living relative remembers as being named

Mayme. They were wed in 1890 and had no children. A granddaughter recalls:

> I remember Grandpop Justin and Grandmom Mayme. She was Catholic and Grandma Barrah had been Irish. Grandpop lived in Pennington. I don't know what he did. Mom used to take us down and we stayed there while Mom went shopping. Grandmom Mayme would give us her big rosary beads with the cross to play with.

In describing Justin and Eunice and commenting on the relationship between criminality and feeblemindedness, Goddard states:

> We have claimed that criminality resulting from feeblemindedness is mainly a matter of environment, yet it must be acknowledged that there are wide differences in temperament and that, while this one branch of the Kallikak family was mentally defective, there was no strong tendency in it towards that which our laws recognize as criminality. In other families there is, without doubt, a much greater tendency to crime, so that the lack of criminals in this particular case, far from detracting from our argument, really strengthens it. It must be recognized that there is much more liability of criminals resulting from mental defectiveness in certain families than in others, probably because of difference in the strengths of some instincts.
>
> This difference in temperament is perhaps nowhere better brought out than in the grandparents of Deborah. The grandfather belonging to the Kallikak family had the temperament and characteristics of that family, which, while they did not lead him into a positive criminality of high degree, nevertheless did make him a bad man of a positive type, a drunkard, a sex pervert, and all that goes to make up a bad character.
>
> On the other hand, his wife and her family were simply stupid, with none of the pronounced tendencies

to evil that were shown in the Kallikak family. They were
not vicious, nor given over to bad practices of any sort.
But they were inefficient, without power to get on in the
world, and they transmitted these qualities to their
descendants. (Goddard, 1912, pp. 62–63)

Goddard then goes on to illustrate once again the inevitable effect of the bad seed:

Thus of the children of this pair, the grandparents of
Deborah, the sons have been active and positive in their
lives, the one being a horse thief, the other a sexual
pervert, having the alcoholic tendency of his father,
while the daughters are quieter and more passive.
Their dullness, however, does not amount to imbecility.
Deborah's mother herself was of a high type moron,
with a certain quality which carried with it an element of
refinement. Her sister was the passive victim of her
father's incestuous practice and later married a normal
man. Another sister was twice married, the first time
through the agency of the good woman who attended to
the legalizing of Deborah's mother's alliances, the last
time, the man, being normal, attended to this himself.
He was old and only wanted a housekeeper, and the
woman, having been strictly raised in an excellent family, was famous as a cook, so this arrangement seemed
to him best. None of these sisters ever objected to the
marriage ceremony when the matter was attended to
for them, but they never seem to have thought of it as
necessary when living with any man. (Goddard, 1912,
pp. 63–64)

It is interesting to contrast Goddard's portrayal of Deborah's maternal aunts and uncles with census and courthouse records and with the recollections of two of Deborah's
half sisters who are still living. Their recollections are cited
below in quotation marks.

Abigail, supposedly feebleminded and the mother of feebleminded children, was born February 26th, 1863. She was listed in the 1870 census as living in her father's, Justin's, home. She married in 1896 and is listed in the census of 1900 as living with her husband and their two children. Her husband's occupation was listed as a laborer, and he was renting a farm at that time. "A third child, a boy, little Dickie died at the age of three months. Aunt Abigail's husband, Uncle Dick, died of rabies about November, 1904. She remarried to Uncle William around 1907." According to the 1910 census, William was a laborer, and he and Abigail were renting a house. There were no children by the second marriage.

Goddard remarked about Gaston that he was "feebleminded and a horse thief [and] he removed to a distant town where he married. He had one child. Mentality of both mother and child undetermined" (Goddard, 1912, p. 29). The recollection of his niece is that "Uncle Gaston lived at Easton. He had a daughter Katherine and a boy Leroy and, I think, a girl Edna. When they came down from Easton for Aunt Ida's funeral I saw Edna and Leroy. They are my second cousins. We got to talking about Mom dying. She and Uncle Gaston both died on April 7th." Records show that actually Katherine was Gaston's only child and that Edna and Leroy were her children.

Not much information was available on Margaret, the supposed victim of Justin's incestuous impulses. Goddard wrote that she

> was taken by a good family when a very small child. When she was about thirteen, she visited her parents for a few weeks. While her mother was away at work, her father, who was a drunken brute, committed incest with her. When the fact became known in her adopted home, she was placed in the almshouse. The child born there soon died, and she was again received into the family where she formerly lived. . . . When about thirty-

five, she married a respectable workingman but has
had no children by him. (Goddard, 1912, p. 28)

Her niece knew only that "Aunt Margaret married a man
named Cochran and lived at Mt. Airy. They never had any
children."

Goddard makes a brief reference to another son of Justin
and Eunice: "Beede, who is feeble-minded. He married a
girl who left him before their child was born. He lives at
present with a very low, immoral woman" (Goddard, 1912,
p. 29).

According to Deborah's half sisters, Beede was a more
complete person than Goddard's description would suggest:
"Yes, Beede was one of Mom's brothers. . . . Beede's wife
Ida died. They had a daughter . . . she lives in Trenton.
Uncle Beede got a job in Lambertville and Aunt Ida wouldn't
go there with him, so he went alone. . . . Pop got Ida and
Uncle Beede back together. Uncle Beede and Ida were Millie's parents. Millie married and had a girl, May, who looked
like Mom." The 1930 city directory of Trenton lists Beede and
Ida and indicates that he was a rubber worker. In 1935, he is
listed as a millworker. Both of them were listed in the 1950
directory. By 1954, Beede had died and Ida was listed as his
widow.

Goddard offers a description of Deborah's mother along
with a bit of a sermon on her irremediable condition in life.
Although parts of this statement were included in chapter 2,
it is important in what it reveals about his perception of the
mother and about his own philosophical and social views
and is thus presented here in complete form:

The stupid helplessness of Deborah's mother in regard
to her own impulses is shown by the facts of her life. Her
first child had for its father a farm hand; the father of the
second and third (twins) was a common laborer on the
railroad. Deborah's father was a young fellow, normal
indeed, but loose in his morals, who, along with others,

kept company with the mother while she was out at service. After Deborah's birth in the almshouse, the mother had been taken with her child into a good family. Even in this guarded position, she was sought out by a feeble-minded man of low habits. Every possible means was employed to separate the pair, but without effect. Her mistress then insisted that they marry, and herself attended to all the details. After Deborah's mother had borne this man two children, the pair went to live on the farm of an unmarried man possessing some property, but little intelligence. The husband was an imbecile who had never provided for his wife. She was still pretty, almost girlish—the farmer was good-looking, and soon the two were openly living together and the husband had left. As the facts became known, there was considerable protest in the neighborhood, but no active steps were taken until two or three children had been born. Finally, a number of leading citizens, headed by the good woman before alluded to, took the matter up in earnest. They found the husband and persuaded him to allow them to get him a divorce. Then they compelled the farmer to marry the woman. He agreed, on condition that the children which were not his should be sent away. It was at this juncture that Deborah was brought to the Training School.

In visiting the mother in her present home and in talking with her over different phases of her past life, several things are evident; there has been no malice in her life nor voluntary reaction against social order, but simply a blind following of impulse which never rose to objective consciousness. Her life has utterly lacked coordination—there has been no reasoning from cause to effect, no learning of any lesson. She has never known shame; in a word, she has never struggled and never suffered. Her husband is a selfish, sullen, penurious person who gives his wife but little money, so that she often resorts to selling soap and other things among her neighbors to have something to spend. At times she works hard in the field as a farm hand, so that it cannot be wondered at that her house is neglected and her children unkempt. Her philosophy of life is the philosophy of the animal. There is no complaining, no irritation at the inequalities of fate. Sickness, pain, childbirth, death—she accepts them all with the same equanimity as she accepts the opportunity of putting a new dress and a gay ribbon on herself and children and going to a Sunday School picnic. There is no rising to the comprehension of the possibilities which life offers or of directing circumstances to a definite, higher end. She has a certain fondness for her children, but is incapable of real solicitude for them. She speaks of those who were placed in homes and is glad to see their pictures, and has a sense of their belonging to her, but it is faint, remote, and in no way bound up with her life. She is utterly helpless to protect her older daughters, now on the verge of womanhood, from the dangers that beset them, or to inculcate in them any ideas which would lead to self-control or to the directing of their lives in an orderly manner. (Goddard, 1912, pp. 64–67)

Martha, the mother of Deborah, was born in April of 1868. According to census data, in 1870 she was living with her parents, Justin and Eunice. Later she lived and served as a domestic and child care helper in the home of a neighbor. She is shown living in this home in the 1885 census records. Deborah was born to Martha out of wedlock.

Martha married her first husband in November of 1889 and divorced him after the birth of two more children. She married her second husband around November of 1897, just after Deborah was admitted to the Vineland Training School. Martha had seven children by her second husband. She died on April 7, 1932. Her second husband died in 1942 and is buried beside her. One of Martha's daughters by her second husband (Deborah's half sister) recalls:

I look like my mother . . . dark eyes and dark hair like Mom. Jenny and Ward look like my father's people. Ward's in Harrisburg. Jess died when she was 60 and Fred drowned when he was in his 40's. Ward didn't do anything. He was the baby. Ward was home and away and then back home again. He fooled around just like a big baby. Mom spoiled him to death—the youngest one. Fred got married. Jess went to stay with Aunt Jane and her husband. Aunt Jane was Pop's sister. Jess went to school there. They spoiled her. She was stubborn— Pop's family were all stubborn.

In the Kallikak book, Goddard comments on Deborah's half brothers and sisters by Martha's second husband:

The last family of half brothers and sisters of Deborah are, at present, living with the mother and her second husband. The oldest three of these are distinctly feeble-minded. Between them and the two younger children there was a stillbirth and a miscarriage. The little ones appear normal and test normal for their ages, but there is good reason to believe that they will develop the same defect as they grow older. (Goddard, 1912, p. 28)

The oldest of these children, one that Goddard described as "distinctly feeble-minded," made the following comments in an interview in May, 1984:

I went to Marshall's Corner school, so did Dot, Jess, and Fred. When trolley cars came along Mom was afraid we'd get run over by one. She told Pop she could just see us tangled up under one and couldn't we go to the mountain school. Pop went to see the trustees and since we lived near the line between the districts—about as far from one school as the other—we started going to the mountain school. I didn't go too far in school. Mom had to help Pop work in the fields and I had to take care of the kids when they came along. I only went to the fifth grade. . . . My teacher, Julia Holcombe, said it was a shame I couldn't come more often because I could have learned. But I did learn how to read and write and figure.

In speaking of her adult life she related:

My daughter Dot was born in March and I was 21 the next July. Eve was born 7 years later. My husband died when Eve was little. I was left with two kids. I could go home to Mom and Pop whenever I needed to. I did practical nursing when babies were born. I got $15 for two weeks and board for me and the girls. If I got a job where I couldn't take the girls, I left them with Mom. I got married again 21 years after my husband died. My second husband was Jewish, Al Katz. I thought we'd grow old together. We were married 20 years. We built this house and he died two years later. He's been dead 30 years last October. My husband was in the Army 31 years so I go to Fort Dix for my doctoring. That's where I had my leg off two years ago. They said I recovered good. I always took care of myself—never smoked or drank.

At the time of the interview, this woman was 86 years old, was about 5 feet 3 inches tall, and weighed about 120 pounds. Except for her amputated leg, she appeared to be in good health. She was lively, alert, and most lucid. Her great grandson Ray is a golf professional. He lives with his family in Florida. "Jenny [her sister] and I go to Florida every year now. I've been going down there for 40 years now. We used to have to pay, but now we stay with Ray and his family."

Jenny was another sibling of Deborah that Goddard considered to be "distinctly feeble-minded." She married a farmer who had a daughter by a previous marriage. Together they had one son, Peter. He served in the Army for

20 years. After he retired, he operated an airport taxi service. He recently retired from that business and moved back close to the area where the Kallikak story originated. According to Jenny, "even though he is retired he spends his week doing lawns and flower beds. He's ambitious and likes to keep busy. . . . [we] are going down to Peter's Thursday for ten days. His wife will come up and get us and we'll stay with them."

The half brother Fred was the third sibling of Deborah from her mother's second marriage that Goddard diagnosed as being defective. This man was married, had children, and was apparently a responsible worker all of his life. He served in World War I and is listed in the Pennington honor roll of veterans. According to his half sister, "Fred had always wanted to be an automobile mechanic. He ended up as boss mechanic at Bob Jones's garage on Main Street for a good many years." The 1930 Trenton city directory lists him as an auto mechanic. Fred is buried beside his mother and father.

In commenting on Deborah's siblings from her mother's first marriage, Goddard remarked:

The next younger half sister of Deborah was placed out by a charitable organization when very young. From their records we learn that in five years she had been tried in thirteen different families and by all found impossible. In one of these she set the barn on fire. When found by our field worker, she had grown to be a girl of twenty, pretty, graceful, but of low mentality. She had already followed the instinct implanted in her by her mother, and was on the point of giving birth to an illegitimate child. She was sent to a hospital. The child died, and then the girl was placed permanently in a home for feeble-minded. An own brother of this girl was placed out in a private family. When a little under sixteen, his foster mother died and her husband married again. Thus the boy was turned adrift. Having been

well trained, and being naturally of an agreeable dis-
position, he easily found employment. Bad company,
however, soon led to his discharge. He has now drifted
into one of our big cities. It requires no prophet to predict
his future. (Goddard, 1912, p. 27)

Here is the recollection of one of Deborah's living half
sisters:

Mom had Deborah first before she was married. She
wasn't Mom's first husband's daughter. She was a mis-
take. Then Mom married and had Harry and Anna. I
don't recall ever seeing either Deborah or Anna. We lost
track of them. Anna was pretty. Mom had a picture of
Anna, she had dark eyes and dark hair. I think I have
that around here somewhere.

The birth records of Mercer County, New Jersey, show that
Harry was born in 1890. He died in 1920. According to his
half sister:

I only remember Harry—he looked like me. He was my
half brother. He died in Donnelly Memorial Hospital in
Trenton of TB. He was only in his thirties. He worked in
the thread mill at Yardville and got what we called
"weaver's consumption." He was in Glen Gardner San-
itarium for about two years and left there to come home.
But the doctors wouldn't allow him to stay at home
because of the children. So he went into the hospital
and only lived about two months. His wife Ella had two
children by her first marriage and she and Harry had
one together.

Anna was born in 1892 in Mercer County and, according
to census listings, lived with her mother and father through
most of her early years. Consistent with Goddard's account
that she was "placed out at an early age," the recollection is

that "Anna went to Nate Blackwell's, I think—I'm not sure. He was the undertaker in Pennington. All the family was buried from there. Mom worked for them too, when she was young."

The 1915 census of Landis Township, Cumberland County, New Jersey, lists the "inmates" of the New Jersey State Institution for the Feeble-Minded. Anna's name appears on the list. The census record indicates that Anna could read, write, and speak English. Why was she in an institution for the feebleminded? Perhaps for the same reason that her sister was living across the street in the training school.

And so we come back to Deborah. The records of the alms-house, or what was then called a "poor farm," show that Martha Kallikak came to the farm in November of 1888. She was admitted by order of the overseers. She gave birth to an illegitimate female child in February of 1889. The child was given the name Deborah. In June of that year the mother and infant left the farm. Deborah lived with her mother until 1897 when she entered the training school at Vineland. Goddard's story of the Kallikaks had its beginning there. The institutions at Vineland would be Deborah's home until her death in 1978. A photograph of Deborah in her late years is shown in Illustration 12.

The story of the disfavored Kallikaks is not free of troubles and human frailties. The family did have its share of illegitimate children, drunkards, "ne'er-do-wells," and the other skeletons that have a way of jumping out of genealogical closets. But so do most families, particularly those who have been faced with poverty, lack of education, and scarce resources for dealing with social change. But the family also had its strengths and successes. The tragedy of the disfavored Kallikaks is that their story was distorted so as to fit an expectation. They were perceived in a way that allowed only their weaknesses and failures to emerge. Their story was first interpreted according to a powerful myth, and then used to bolster further that myth. The myth was that of

Illustration 12: Deborah Kallikak at age 73. Reprinted with permission from *Charity and Corrections in New Jersey: A History of State Welfare Institutions* by J. Leiby, 1967, New Brunswick, N.J.: Rutgers University Press. Courtesy of Vineland Development Center, Vineland, N.J.

eugenics. All the "bad" Kallikaks were bad, and that legacy would remain unchanged.

While I was doing research in New Jersey during the late winter of 1983–1984, a local newspaper carried two articles on a young woman from the area who had distinguished herself academically and through extracurricular activities at a respected midwestern college. She was an honor student and had been recognized for her artistic talent. This outstanding young person is the great, great, great granddaughter of Martin Kallikak, Jr., through his daughter, "Old Sal"—the most recent flowering of the bad seed.

REFERENCE

Goddard, H.H. (1912). *The Kallikak family: A study in the heredity of feeble-mindedness.* New York: MacMillan.

Chapter 8

Immigrants, Morons, and Democracy

B Y THE BEGINNING OF THE TWENTIETH CENTURY, PATTERNS OF immigration to the United States had undergone dramatic changes. The "old" immigrants had been for the most part Anglo-Saxons, Germans, and Scandinavians. They were welcomed to America as poor but good stock who would add strength and vigor to the society. These were the "huddled masses yearning to breathe free" of the poem by Emma Lazarus inscribed on the Statue of Liberty. But by 1900, most of the immigrants to the United States were of other stocks. They were Italians, Poles, Hungarians, Slavs, and Russians; they were "different." Americans of the time, forgetting their own immigrant heritage, began to view the newcomers in less romantic terms. They were more likely to think of the immigrants from eastern and southern Europe as, in the words of a less quoted line from Lazarus's poem, the "wretched refuse" of Europe's "teeming shores." Europe's human trash was seen to be washing into New York harbor.

Goddard had created the term moron to explain social ills. Morons, the "high-grade" feebleminded, were more difficult to detect than seriously retarded people and therefore posed a great threat to society. As the primary source of crime, poverty, alcoholism, and sexual irresponsibility, they had to be detected, segregated, and prevented from reproducing. The Kallikak story was a testament to the menace of the moron to the United States.

Thus, Goddard soon turned his attention to the question of immigration. He was invited to Ellis Island to observe the procedures used in processing the immigrants and determining their "fitness" to enter the United States. On the basis of his observations, he was asked to make suggestions as to how immigrants could be examined more effectively for the purpose of detecting mental defectives. Goddard came away with the impression that the physicians on Ellis Island were simply looking the immigrants over for obvious physical signs of mental deficiency. But Goddard had argued earlier that most morons, the real threat to society, look no different from other people. In describing his first visit to the island, he remarked:

> Both Professor Johnstone and myself were much discouraged. We went home and said we didn't see how much could be done because of the great number that were coming in every day. There are about 5,000 a day, 29,000 in a week, and there are comparatively small facilities for handling them. I went again last spring, and was able to look a little more intelligently at it. (Goddard, 1917, p. 105)

Goddard decided that the problem of detecting immigrant morons could be approached by employing some of the same techniques he had used in the Kallikak study. First, he would use field workers, women he had trained at Vineland in "moron detection," who would be able to recognize feebleminded immigrants by simple visual inspection. Second, he would use standardized intelligence tests (primarily the Binet) to determine the mental ages of various immigrant groups.

Goddard's confidence in the ability of trained persons to spot feebleminded people by noting subtle differences in appearances was remarkable. In justifying the use of this technique, he commented:

After a person has had considerable experience in this work, he almost gets a sense of what a feeble-minded person is so that he can tell one afar off. The people who are best at this work and who I believe should do this work, are women. Women seem to have closer observation than men. It was quite impossible for others to see how these two young women could pick out the feeble-minded without the aid of the Binet test at all. (Goddard, 1917a, p. 106)

Goddard describes placing a young woman at the end of a line of people waiting to be processed. As the immigrants passed her, she pointed out the ones she thought were morons. These people were taken to a separate room and tested with the Binet. Goddard boasts that the woman picked out nine people whom she judged to be defective and that, according to the Binet test, every one of the nine was below normal.

Goddard was confident that the Binet tests that he used to confirm his field workers' impressions were equally valid when used with newly arrived immigrants. It is interesting to note that Goddard expressed some concern about using the Binet through an interpreter. He felt that interpreters were inclined to prompt or encourage the immigrant and thereby bias the results. He was also concerned that the interpreter might not correctly translate the Binet.

Apparently, however, Goddard largely disregarded the effects of language difficulties, fatigue from the long ocean voyage, fear, and cultural differences when he interpreted the Binet test scores of immigrants. He describes one immigrant's test performance:

> We picked out a young man whom we suspected was defective, and, through the interpreter, proceeded to give him the test. The boy tested eight by the Binet scale. The interpreter said, "I could not have done that when I came to this country," and seemed to think the test unfair. We convinced him that the boy was defective. (Goddard, 1917a, p. 105)

In the spring of 1913, Goddard was provided with funds to support a study of newly arrived immigrants based on the observation of physical traits and mental testing. He sent members of his Vineland staff to Ellis Island where they spent two and a half months screening arriving immigrants and testing those they suspected of being feebleminded.

Goddard's workers selected for testing some immigrants whom they thought to be feebleminded (these were chosen from groups of Russians and Italians being processed on the island). In order to assess the intelligence of "average" immigrants, the workers also picked out people who appeared to them to be representative of the various groups arriving at that time (Jews, Hungarians, Italians, and Russians).

In explaining this selection procedure, Goddard stated:

> In both instances the cases were selected after the government physicians had culled out all mental defectives that they recognized as such. On the other hand the very obviously high grade intelligent immigrant was not selected. Our study therefore makes no attempt to

determine the percentage of feeble-minded among immigrants in general or even of the special groups named—the Jews, Hungarians, Italians and Russians. At the same time it must be remembered that these superior individuals, who were not included in our study, were so small a part of the group that they did not noticeably affect the character of the group. As stated the physicians had picked out the obviously feeble-minded, and to balance this we passed by the obviously normal. That left us the great mass of "average immigrants." (Goddard, 1917b, p. 244)

So, though inserting a disclaimer and acknowledging that the selected samples of immigrants were not representative of the groups they were picked from, Goddard believed that they really were typical of the immigrants arriving at that time.

The results of the Binet testing of the immigrant samples is incredible in more ways than one. As shown below in the figures reproduced from Goddard's report, over 83 percent of all the Jews tested were feebleminded, as were 80 percent of the Hungarians, 79 percent of the Italians, and 87 percent

Intelligence Classification of Immigrants of Different Nationalities

	Normal		Borderline		Feebleminded		Moron		Imbecile	
	No.	%	No.	%	No.	%	No.	%	No.	%
Jews	3	10	2	7−	25	83+	23	76	2	7
Hungarians	0	0	4	20	16	80	16	80	0	0
Italians	3	7−	7	15−	38	79	38	79	0	0
Russians	0	0	4	9	39	87	37	82	2	2.5
Italian F.M.	0	0	1	5+	17	94+	12	63	6	32−
Russian F.M.	0	0	0	0	18	100	14	78−	4	22+

Note: The plus and minus signs following percentage figures apparently indicate that the actual numbers are slightly higher or lower than those reported. The percentages given for morons and imbeciles are a breakdown of the total percentage for feebleminded.

(Reprinted with permission from "Mental Tests and the Immigrant" by H.H. Goddard, 1917, *Journal of Delinquency, 2,* p. 252.)

of the Russians (it is interesting to note that Jews were grouped and tested according to religion rather than nationality). He provides comparison figures of percentages of Italians and Russians diagnosed as feebleminded by field worker judgment, or what he calls his "first method" (F.M.), which are even higher.

Goddard himself recognized how difficult it was to believe the magnitude of these percentage figures. He made some adjustments and invalidated the test scores for some individuals but still found that around 50 percent of all of the immigrants tested were feebleminded. He comments:

> Doubtless the thought in every reader's mind is the same as in ours, that it is impossible that half of such a group of immigrants could be feeble-minded, but we know that it is never wise to discard a scientific result because of apparent absurdity. Many a scientific discovery has seemed at first glance absurd. We can only arrive at the truth by fairly and conscientiously analyzing the data. (Goddard, 1917b, p. 266)

A sense of the real absurdity of such mental testing—and of the inordinate confidence Goddard had in the results—can be derived from an examination of the raw data included in the report. As an example of how the information was recorded and analyzed by Goddard and his field workers, the report gives test scores for the Jewish group. By focusing on information available on a few of these people, we can see how the test scores were regarded as much more indicative of true mental ability than the actualities of the people's lives (see the data reproduced below from Goddard's report).

Subject 1 is a 21-year-old man with only five years of education. His limited education might well have been due to limited opportunities and economic necessity. Whatever the case, we would assume that his chances for academic learning had not been extensive. He scores a mental age of

Data on Jewish Immigrants as Recorded by Goddard's Field Workers

Subject	Sex	Age	Mental Age (Binet)	School Experience	Remarks
1	M	21	12–1	5 years	Tailor, knows three languages
2	M	21	11–4	Until 13	Tailor
3	M	21	11–3	6 years	Apothecary's assistant. Diploma
4	M	23	11–1	Until 14	Works in leather
5	M	19	11	Synagogue until 13	Tailor
6	M	17	10–3	Synagogue	No occupation. Writes a little.
7	M	20	10–3	5 years	Tin worker. Idiotic look.
8	F	14	10–3	Until 13, 7th grade	
9	M	22	10–2	Only synagogue	Telegrapher
10	F	16	10–2	None	Did not work in Russia
11	M	15	10–2		Grocery clerk
12	F	20	10–1	4 years, 4th class	Tailor
13	M	21	10–1		Held as defective. Then discharged
14	M	19	10–1	Synagogue only	Detained 1 night. Waiter in Poland
15	M	22	10–1	Synagogue only, until 13	Printer
16	M	21	10–1	Synagogue only	Tailor
17	M	20	10	Synagogue 6 years	Seltzer works. Insane?
18	M	19	9–9	3 years	Farmer
19	M	17	9–4	Synagogue, until 12 years	Tailor. Read & Write. Russian & Jewish
20	F	22	9–3	None	
21	M	17	9–0	2 years synagogue	Tailor
22	M	17	9–3	5 years	Farmer. Learned Polish & Russian
23	M	18	9–3	Only synagogue	
24	M	20	9–2	Synagogue until 10	Tailor
25	F	11	9–2	Only synagogue	Held overnight. Epileptic.
26	M	18	8–4	Only synagogue	Baker
27	M	24	8–4	Only synagogue, until 12	Tailor
28	F	9	8–2	1 year	
29	M	10	8–1	None	
30	M	9	8	Only synagogue	
31	M	24	7–4	Only synagogue, 7 years	Merchant in farm produce
32	F	11	7–2		Borderline
33	F	18	7	None	

(Adapted with permission from "Mental Tests and the Immigrant" by H.H. Goddard, 1917, *Journal of Delinquency*, 2, p. 246).

12 years, 1 month on the Binet. He is thereby classified as a moron. When we glance at the remarks column, however, we see that the field worker has noted that the man is a tailor and that he knows three languages. This information would seemingly contradict an image of mental defectiveness. Yet the belief in the validity of the test over other information was apparently so strong that there was no hesitation in recording what appears to be an absurd contradiction, and no need to explain it was perceived.

Subject 3 is a 21-year-old man who has a diploma and has worked as a pharmacist's assistant. His mental age on the test is 11 years, 3 months. He is listed as a moron.

Subject 9 is 22 years of age. He is a telegrapher whose only schooling was in the synagogue. He scores a mental age of 10 years, 2 months on the Binet. He is recorded as a moron.

The absurdity and tragedy of classifying people on a single abstract measure becomes clear when we look at individual cases like these rather than statistical summaries about groups of people. The fact that tragic generalizations continue to be made about groups of people based on such measures is discussed later in the book.

Calling on his experience in the Kallikak study, Goddard intended to follow up on the "feebleminded" immigrants who were detected on Ellis Island. He planned to send his field workers out to find the immigrants after a year or two to confirm that they were indeed morons and that they were encountering and causing the social problems he viewed as characteristic of that group. That project turned out to be unsuccessful for the most part because of the difficulty in locating the immigrants after an extended period of time. The field workers found that the immigrants had often changed or anglicized their names or were living at addresses where the household was listed under someone else's name, and they found that the language barrier in ethnic neighborhoods made it difficult to locate their subjects. The committed and enthusiastic Elizabeth Kite, however, appears in Goddard's account of the research with information about one of the immigrants she had been assigned to find:

> When seen the girl was neat in appearance, seemed to be on good terms with the other servants, and though rather dull and stolid-looking answered the questions put to her fairly well. The reasons she gave for deciding to come to America were childish. It seemed to have been a matter of "just taking a notion to" and then coming. Miss Kite, our investigator, reports: "On September 25th, 1916, I called again at the Academy. The housekeeper was not home but I saw two over-servants who told me R. had left them six months ago. . . . The lady of the house was not at home but I questioned the cook, an intelligent Irish woman, who gave a good report of R., but as I questioned further she offered to find the mother of the lady of the house, who was upstairs. She came down and I found her sufficiently intelligent to go into a quite thorough analysis of R.'s mentality as shown in her work and care of the children. She said R. had stayed with them about six months and had been in most ways satisfactory. She was perfectly honest, reliable and industrious, neat and good to the children.
>
> Why then was she discharged? . . . There was about this girl a certain obstinacy, a determination to do her own way, of which they had been told when she came to them. This peculiar mental state seemed incurable. The mistress had often been annoyed by it and finally the outbreak came.
>
> There seems little doubt that this mental state has directly to do with intelligence and comes from a certain lack of power of comprehension, but apart from this I could get no history of anything bordering on what we know to be characteristic of feeble-mindedness." (Goddard, 1917b, p. 265)

Shades of Deborah Kallikak! Only the lack of more information seems to have prevented a definite diagnosis in R's
case.

In his discussion of the results of the research, Goddard
made statements that were very uncharacteristic of him. Up
to that point in his career, he had insisted that morons were
the result of heredity and that the best solution to the problems they posed was segregation by institutionalization. In
his 1917 report, he suggests the possibility that the high rate
of incidence of morons among immigrant groups may be the
result of the environment, and he argues that there may be a
place for them in American society. Whether this change
came about because of the weight of the numbers of immigrants classified as morons (it would be impossible to institutionalize 50 percent of the incoming population) or because
of other factors is not clear. In any case, he would continue to
develop this line of thought in the years to come. His conversion, however, was not complete nor as dramatic as it might
have at first appeared. This becomes apparent in our later
discussion of his political views. It should also be remembered that Goddard never changed his stand on the Kallikak
study, its meaning, and its validity. He continued throughout his life to support the hereditary view of most retardation.

On the question of the immigrant moron, Goddard asks:

> Are these immigrants of low mentality cases of heredi-
> tary defect or cases of apparent mental defect by depri-
> vation? . . . We know of no data on this point, but
> indirectly we may argue that it is far more probable that
> their condition is due to environment than that it is due to
> heredity. To mention only two considerations: First, we
> know their environment has been poor. It seems able to
> account for the result. Second, this kind of immigration
> has been going on for 20 years. If the conditions were
> due to hereditary feeble-mindedness we should prop-
> erly expect a noticeable increase in the proportion of the

feeble-minded of foreign ancestry. This is not the case. (Goddard, 1917b, p. 270)

Speaking of the possible role of these immigrants in society, Goddard begins to view them as social and economic instruments:

It is perfectly true there is an immense amount of drudgery to be done, an immense amount of work for which we do not wish to pay enough to secure more intelligent workers. It is a very big social and economical problem and one which we cannot at this time discuss, as to what kind of adjustment or arrangement society ought to make for getting this work done. May it be that possibly the moron has his place? (Goddard, 1917b, p. 269)

Although Goddard may have softened his stand on the issues of feeblemindedness and institutionalization, he left the basic questions surrounding immigration open to study and action. He concludes his report by saying, "All of this means that if the American public wishes feeble-minded aliens excluded, it must demand that Congress provide the necessary facilities at the ports of entry" (Goddard, 1917b, p. 271).

By facilities, Goddard meant centers for mental testing that would detect feeblemindedness not diagnosed through medical examination. It is also important to note that in his conclusion Goddard reported with obvious pride that "beginning at about the time of our experiment, the number of aliens deported because of feeble-mindedness . . . increased approximately 350 percent in 1913 and 570 percent in 1914 over what it had been in each of the five preceding years" (p. 271). He attributed these increases to the work of the physicians who became inspired by their belief in the use of mental tests for the detection of morons.

During the summer of 1917, Henry Goddard participated in another project that was to have far-reaching social consequences. He was invited by Robert Yerkes (a fellow psychologist who also was interested in mental testing and the heredity of intelligence) to help in the design and construction of mental tests for the U.S. Army; the tests were to be used with recruits during World War I. Goddard, Yerkes, Lewis Terman, and several others worked on the tests at the training school in Vineland. The results were the Army Alpha and Beta. The Alpha was in written form for use with literate recruits, and the Beta used pictures to test those who could not read. More than a million and a half recruits were given these tests during the course of World War I.

The results of this massive screening of American men were published in several army reports beginning in 1918. On the basis of these reports, Goddard began to argue that the average intellectual level in the population was a mental age of 12. For an adult male, a mental age of 12 years placed him within the moron classification. Goddard argued that the results of the army testing showed "beyond dispute" that "half the human race exists at a level little above the moron" (Goddard, 1919, p. 234). Later he became more specific; he presented data from the army testing indicating that 45 percent of all the recruits tested had mental ages below 13 years. On this basis, he projected that 45 percent of the entire population, if tested, could be classified as morons. He concluded that, even though the average American adult had only the mentality of a 13-year-old child, almost half were even less intelligent. He felt that the prospects for universal education for all American children was poor, since 45 percent did not have the capacity to go beyond elementary school and 70 percent could not go beyond the eighth grade (Goddard, 1920).

Goddard's interpretations of the Alpha and Beta test results, together with those of Terman and others, created considerable concern about the state and possible decline of the "American intellect." In 1923, C.C. Brigham of Princeton, following up on these reports, examined the army test results in his provocative and influential book, *The Study of*

American Intelligence. He examined the results in relation to immigration and concluded that immigrant recruits scored lower than American-born recruits and that immigrants from southern and eastern Europe scored lower than those from northern Europe. He argued that continued immigration from the south and east of Europe posed a serious threat to the level of the American intellect and that immigration from these areas should be stopped (Brigham, 1923).

Walter Lippmann, the journalist, was probably the strongest critic of these interpretations of the army test data. He believed that the intellectual and theoretical biases of the psychologists who were making these interpretations had blinded them to the differences in the environmental backgrounds of the men taking the tests and to the influence of these differences on their test scores. He was particularly incensed by Brigham's book. He accused Brigham and other psychologists who were making hereditarian interpretations of the test results of "offering yellow science to the public" (Pastore, 1978).

The Immigration Restriction Act was passed in 1924. Some scholars have questioned the degree to which Goddard's research on immigrants and his interpretation of the army test results influenced the passage of that legislation (Snyderman & Herrnstein, 1983). It is known that the army data were quoted in congressional debate on the issue and that Brigham's book was submitted to the Committee on Immigration during hearings on the act. More important, however, was the impact of Goddard's work in shaping public opinion about immigration.

In its final form, the Immigration Act placed the heaviest restrictions on eastern and southern Europeans. The national groups that Goddard had found to be filled with feeblemindedness—Italians, Russians, Hungarians, and Jews from all over eastern Europe—were among those who were no longer welcome in the United States.

Stephen Jay Gould eloquently portrays the impact of the restrictions:

> The quotas stood, and slowed immigration from southern Europe to a trickle. Throughout the 1930s, Jewish refugees, anticipating the holocaust, sought to emigrate, but were not admitted. The legal quotas, and continuing eugenical propaganda, barred them even in years when inflated quotas for western and northern European nations were not filled. . . . We know what happened to many who wished to leave but had nowhere to go. The paths to destruction are often indirect, but ideas can be agents as sure as guns and bombs. (Gould, 1981, p. 233)

In his book, *Psychology of the Normal and Subnormal,* published in 1919, Goddard explored the political implications of his work on feeblemindedness. He discussed the indications from the testing of draftees that the average intelligence was that of a 13-year-old and questioned whether democracy could truly exist with such a populace. He commented:

> It certainly is an argument against certain theories of democracy. . . . To maintain that mediocre or average intelligence should decide what is best for a group of people in their struggle for existence is manifestly absurd. We need the advice of the highest intelligence of the group, not the average, any more than the lowest. (Goddard, 1919, p. 236)

Democracy, according to Goddard, means that people select the wisest and most intelligent to "tell them what to do to be happy. Thus Democracy is a method for arriving at a truly benevolent aristocracy" (Goddard, 1919, p. 237). He felt that the truest democracy he had observed was in an institution for the feebleminded. In a speech at Princeton University, he said:

The inmates of the Vineland Training School, imbeciles and morons, did not elect Superintendent Johnstone and his associates to rule over them; but they would do so if given a chance because they know that the one purpose of that group of officials is to make the children happy. (Goddard, 1920, pp. 98–99)

In some ways, Goddard's vision was that society should function more like an institution. The army data should prepare us to accept that there are many more morons in society than previously thought, almost equal in numbers to the normal population. He said that recognition of this fact

will prepare us to accept the findings . . . showing that large groups of so-called menials really fall into the moron class. This again enables us to understand their shortcomings, their follies, their blunders and failures. And, what is still more important, it points the way to a wise and satisfactory treatment of these classes by the more intelligent group. (Goddard, 1919, p. 238)

In discussing the management of the vast numbers of feebleminded people that he had come to believe existed in society, Goddard was pessimistic about the efficacy of education. He felt that "lives and fortunes" had been spent attempting to educate these people to no avail. He felt that special methods, special schools, and instruction in institutions had failed to make a difference in the ability of morons to function more effectively in society. His solution to the problem was, once again, to see that they were controlled and treated compassionately by the intellectual aristocracy.

In speaking to a group of Princeton students, Goddard continued his analysis of the relationship between mental ability and participation in a democracy:

Whenever the four million choose to devote their superior intelligence to understanding the lower mental lev-

els and to the problem of the comfort and happiness of
the other ninety-six million, they will be elected the
rulers of the realm and then will come perfect government—Aristocracy in Democracy. . . . While we all
believe in democracy, we may nevertheless admit that
we have been too free with the franchise and it would
seem a self-evident fact that the feeble-minded should
not be allowed to take part in civic affairs; should not be
allowed to vote. (Goddard, 1920, p. 99)

It is important to note that the implication of this statement is
that nearly half of the adult population should be disen-
franchised and lose the right to vote.

Goddard told the students that other ideas about equality
were equally ridiculous. For example, he said that appeals
for equality in housing were as absurd as insisting that every
laborer receive a graduate fellowship. He felt that people of
differing levels of intelligence not only had different capaci-
ties but also different requirements for happiness. Goddard
emphasized that mental levels are fixed and cannot be
altered by education or changes in circumstance:

Much money has been wasted and is continually being
wasted by would-be philanthropists who give liberally
for alleviating conditions that are to them intolera-
ble. . . . They do not understand that it is being wasted
because the people who receive it have not sufficient
intelligence to appreciate it and use it wisely. Moreover,
it is a positive fact that many of these people are better
contented in their present surroundings than in any that
the philanthropists can provide for them. (Goddard,
1920, p. 103)

By 1928, Goddard had apparently softened his views on
the fixed nature of feeblemindedness and the necessity of
institutionalizing morons. Writing in the *Journal of Psycho-*

Asthenics, he said that he had reached a point where he no longer believed in the incurability of the moron. He felt that the right kind of education and training could prepare the moron to function better in society. Although intelligence could not be raised (a "poor" brain would still be a "poor" brain), the moron could be taught to use what he had more effectively. His opinion on the necessity of segregating morons and preventing marriage and reproduction among them had taken a dramatic turn. He had come to believe that, since most people of inferior intellect married and had children, those identified as morons should not be denied that privilege. In the same vein, he noted that if "moronity is only a problem of education and the right kind of education can make out of them happy and useful hewers of wood and drawers of water, what more do we want?" (Goddard, 1928, p. 223).

But Henry Goddard's conversion once again was far from total or permanent. Four years later, in the anniversary address at the Vineland Training School, his remarks had the flavor of his preconversion views:

> A few years ago William Allen White, of Kansas, startled the country by an article in *Collier's Weekly* in which he asked the question, "What is the matter with America?" His answer, in brief, was "the moron majority." In the main he was right . . . we know now that the so-called civilized nations are made up of people of a wide range of capacity and, consequently, of responsibility. One half of the world has not the intelligence, the capacity, to become even interested in the great social problems. It requires the entire mental energy of these people to get enough to eat, clothes to keep them warm and automobiles to transport them. . . . It is perfectly clear to those who understand this situation that half of the world must take care of the other half. (Cited in Doll, 1932, pp. 58–59)

In his address, Goddard also reaffirmed the importance of the army mental testing project and endorsed the interpretation he had made of the results:

> And so I repeat that, in my judgment, the knowledge derived from the testing of the 1,700,000 men in the Army is probably the most valuable piece of information which mankind has ever acquired about itself. The information has been hard to accept. We could not believe it. It was thought that the tests were wrong, but more than a dozen years of experience, criticism and testing the tests has strengthened their validity rather than weakened it. (Goddard, 1932, p. 59)

In 1949, *The Nature-Nurture Controversy* by Nicholas Pastore presented an analysis of the views of a select group of prominent social scientists on the nature-nurture issue and their attitudes on social, political, and economic issues. The chapter on Goddard examined his study of the Kallikaks and his later work. As would be expected, Pastore classified Goddard as a hereditarian on the nature-nurture question; he categorized him as a conservative on social and political issues. Pastore found that most hereditarians were politically conservative (Pastore, 1949).

During the summer of 1983, I found a draft of Pastore's chapter on Goddard in the Goddard papers in the Archives of the History of American Psychology at the University of Akron. I discovered that the book was first written as a dissertation by Pastore while he was a student at Teachers College, Columbia University. The dissertation was sponsored by Professor Goodwin Watson. Pastore had apparently sent the draft to Goddard for his critique and comments.

In Goddard's papers, there is a copy of a letter that he sent to Pastore in response to the draft. In the letter, dated April 3, 1948, Goddard expresses his dissatisfaction and disturbance from his reading of the manuscript. He seems to resent the way in which he was characterized relative to the nature-nurture issue and even denies that the controversy existed at the time of the Kallikak study. We should probably keep in mind that Goddard was 82 years old when he wrote the letter. I mention this because of anecdotal comments I have heard indicating that he was ill during his later years. Here are some excerpts from the 1948 letter:

> I think perhaps you may have been misled by having the answer, before you had the problem. . . . Did you realize that my work was done some forty years ago when the problem of nature-nurture was not formulated? . . .
>
> It is perfectly natural that you should interpret my language in terms of today's experiences. But unfortunately that does not give you *facts* as much as it gives you what *you judge* to be the facts, or even what you *wish* had been the facts; we all do much wishful thinking. . . .
>
> [I] was NOT led to "emphasize heredity and deemphasize environment." [I] was studying heredity and had no inclination to deemphasize environment, because in those days environment was not being considered. . . . A little farther on you fall into the error of perhaps hasty writing thereby misrepresenting [my words]. It is not the criminal that is unmodifiable. Much can be done for the criminal and the pauper. It is their inferior brain which they have inherited that cannot be modified. . . .
>
> The defective brain cannot be changed anymore than an inherited absence of eyes. (Goddard, 1948, p. 6)

Finally, in questioning Pastore's ability to comprehend how he executed his work and made his interpretations, Goddard includes a statement reminiscent of those he had made earlier in his career: "The feeble-minded are very difficult to understand. Like so many other problems, it is the expert, the person who has lived among a group of known feeble-minded, who gets to know them." (Goddard, 1948, p. 6)

REFERENCES

Brigham, C.C. (1923). *A study of American intelligence.* Princeton, N.J.: Princeton University Press.

Doll, E., Ed. (1932). *Twenty-five years: A memorial volume in commemoration of the twenty-fifth anniversary of the Vineland laboratory.* Vineland, N.J.: The Training School.

Goddard, H.H. (1917a). The Binet tests in relation to immigration. *Journal of Psycho-Asthenics, 2,* 105–107.

Goddard, H.H. (1917b). Mental tests and the immigrant. *Journal of Delinquency, 2,* 243–277.

Goddard, H.H. (1919). *Psychology of the normal and subnormal.* New York: Dodd, Mead and Company.

Goddard, H.H. (1920). *Human efficiency and levels of intelligence.* Princeton, N.J.: Princeton University Press.

Goddard, H.H. (1928). Feeble-mindedness: A question of definition. *Journal of Psycho-Asthenics, 33,* 219–227.

Goddard, H.H. (1948). [Letter]. *Goddard papers* (Box M32, Ephemiris). Akron, Ohio: University of Akron, Bierce Library, Archives of the History of American Psychology.

Gould, S.J. (1981). *The mismeasure of man.* New York: W.W. Norton.

Pastore, N. (1949). *The nature-nurture controversy.* New York: King's Crown Press, Columbia University.

Pastore, N. (1978). The army intelligence tests and Walter Lippmann. *Journal of the History of the Behavioral Sciences, 14,* 316–327.

Snydermann, M., & Herrnstein, R. (1983). Intelligence tests and the Immigration Act of 1924. *American Psychologist, 39,* 986–995.

Chapter 9

Eugenics, Sterilization, and the Final Solution

G ODDARD'S STUDY OF THE KALLIKAKS NOT ONLY LED HIM TO find feeblemindedness in various quarters, it also stimulated others to search for evidence that would confirm that the existence of the moron was the basis of most social problems. While Goddard was testing immigrants, mental defectives were being discovered with amazing frequency in America's jails and slums. In Boston, a study of criminals found that less than 8 percent had normal intelligence; a Kansas prison reported that over 68 percent of the white inmates and more than 90 percent of the black inmates were morons; a jail in Virginia found that 64 percent of the inebri-

135

ates and criminals arrested were feebleminded. Consistent with Goddard's portrayal of the sexual misbehavior of the bad Kallikaks, a study in Albany, New York, indicated that at least 85 percent of the local prostitutes were feebleminded. A Pennsylvania field worker found that 98 percent of the unwed mothers she studied were morons (Wallin, 1917).

Henry Goddard created not only the term but also the concept of the moron. His work on the Kallikak family, and later on immigrants and draftees, fortified that concept. The moron as a menacing social sore became a powerful and pervasive image. The idea that a multitude of social problems could be attributed to a single source, feeblemindedness, was most appealing. The concept of the menacing moron evolved into a corpus of conventional wisdom. That moronity was usually hereditary and could not be cured was widely accepted (this remained true even after Goddard had modified some of his own views). The idea that intelligence tests could be used to fairly and accurately diagnose feeblemindedness was rarely challenged. The argument that morons were the primary wellspring of social problems gained widespread popularity, even among social workers and social reform groups. The proposal that these people should be segregated and that their reproduction should be prohibited came to be seen as a reasonable solution to the problems they created.

The result was a proliferation of social policy recommendations based on the belief that the moron constituted the primary source of ills of society. In their book, *Applied Eugenics*, Popenoe and Johnson suggested a number of social reforms. Compulsory education and restriction of child labor were advocated, based not on the humanitarian and equalitarian principles with which they had become associated but on the assumption that these changes would make children more expensive for parents. This would in turn have the effect of pressuring the poorer classes to have smaller families. Fewer children among the poor, an obviously inferior class of people, would then reduce the burdens they placed on society. For the same reason, Popenoe and Johnson opposed such aid to poor children as free school lunches and textbook subsidies. Such programs, they argued, would lower the cost of children to poor families and encourage irresponsible reproduction (Popenoe & Johnson, 1918).

In reading the eugenic literature produced in the first four decades of the twentieth century, one is often struck by the zeal and passion of the writing. Eugenics had the spirit of a religious cause or a political reform movement. The eugenics enthusiasts pushed vigorously for government action that would protect and promote the overall good of society. In the process, some individual liberties might have to be restricted, but the destiny of the nation must come foremost.

Goddard's conception of the moron, the menace of the feebleminded, would thrive and grow for many years. Eventually, society would come to view the sacrifice of the rights of those considered defective for the benefit of the culture—for the preservation of better "human stock "—to be not only allowable but desirable.

Initially, as we have noted, Goddard felt the problem of moronity could best be managed by segregation and the prevention of reproduction. The technique he proposed to accomplish this was institutionalization. Through separation and control, he felt that the number of feebleminded people in the population would be dramatically reduced in a single generation and could practically be eliminated eventually. He considered sterilization to be a less desirable solution.

Goddard's colleagues at the Eugenics Record Office, however, were enthusiastic advocates of sterilization. They lectured and wrote in favor of sterilization on the basis of the hereditary research that had been done by their office, by Goddard, and by other eugenicists. The most zealous of these supporters of compulsory sterilization was Harry H.

Laughlin. He was brought to the Eugenics Record Office by Charles Davenport, the founder, in 1917. Soon Laughlin became deeply committed to the movement to pass state laws requiring sterilization of people judged to be hereditary defectives. In this category he included tramps, beggars, alcoholics, criminals, the feebleminded, the insane, epileptics, the physically deformed, the blind, and the deaf. It is interesting to note that Laughlin himself was epileptic. He was married but had no children. Whether the latter situation was voluntary or not, I do not know.

During the 1920s, Laughlin widened his interest to include the issues of race and immigration. In 1920, he appeared before the House Committee on Immigration and Naturalization. There he testified that immigrants from the eastern and southern parts of Europe were disproportionately feebleminded and were therefore contributing inordinately to the social problems of the country. He expressed his strong concern that they were a threat to the quality of the American stock. He was subsequently appointed as the committee's "expert eugenics agent" (Voorhees, 1981).

Laughlin developed a model for sterilization laws that he presented to state governments as well as to foreign governments. In developing this model, he hoped to influence

law-makers who have to decide upon matters of policy to be worked out in legislation regulating eugenical sterilization; . . . judges of the courts upon whom, in most states having sterilization statutes, devolves the duty of deciding upon the constitutionality of new statutes, and of determining cacogenic [genetically defective] individuals and of ordering their sexual sterilization; . . . administrative officers who represent the state in locating, and in eugenically analyzing persons alleged to be cacogenic, and who are responsible for carrying out the orders of the courts; . . . individual citizens who, in the exercise of their civic rights and duties, desire to take the initiative in reporting for official determination and action, specific cases of obvious family degeneracy. (Laughlin, 1922, p. vii)

Obviously Laughlin had laid out some very broad goals. In addition to influencing public officials and professionals, he hoped to enlist ordinary citizens in advancing the cause of sterilization by reporting their neighbors. That last statement in the above extract, obviously made with no reservation, has an ominous ring to it.

By 1938, more than 27,000 compulsory sterilizations had been performed in the United States (Marks, 1981). Thirty of the state governments had passed sterilization laws, most of them influenced by Laughlin's lobbying efforts and many based on his model law. Although the first sterilization law was enacted in Indiana in 1907, the constitutionality of compulsory sterilization was not fully tested until 1927. In that year, the Supreme Court upheld the right of a state to enforce sterilization against the will of an individual judged to be defective. The case, *Buck v. Bell* (1927), involved the state of Virginia and Carrie Buck, a young woman who had been committed to the State Colony for Epileptics and Feeble-Minded near Lynchburg. Carrie, an 18-year-old Caucasian girl, had been chosen as the first person to be sterilized under Virginia's new law. The right of the state to perform the operation was challenged, and the case was taken to the Circuit Court of Amherst County. The Eugenics Record Office sent a field worker, Arthur H. Estabrook, to assist the state by collecting information on Carrie's heredity. This information was analyzed by Laughlin, who then presented his findings to the court in support of the state's sterilization law (Chase, 1977).

In his statement to the court, Laughlin testified that Carrie was feebleminded, as

evidenced by failure of mental development, having a chronological age of 18 years, with a mental age of 9 years, according to Stanford Revision of Binet-Simon

Test, and of social and economic inadequacy; has a record during her life of immorality, prostitution, and untruthfulness; has never been self-sustaining; has had one illegitimate child, now about six months old and supposed to be mental defective." (Laughlin, 1929, p. 16)

In his account of Carrie's mother, Emma, Laughlin presented a test score indicating that the 52-year-old woman had a mental age of only 7 years, 11 months, and that she was socially inadequate. He described her as immoral, untruthful, maritally unworthy, and a prostitute. He asserted that she had been divorced by her husband on the grounds of infidelity.

In commenting on Carrie Buck's heritage, Laughlin said, "These people belong to the shiftless ignorant, and worthless class of anti-social whites of the South . . . [they are an] ignorant and moving class of people, and it is impossible to get intelligent and satisfactory data" (Laughlin, 1929, p. 17).

Even though he complained of the difficulty of acquiring satisfactory data on the family, Laughlin apparently felt he had enough evidence to testify that Carrie Buck was feeble-minded because of genetic factors:

Generally feeble-mindedness is caused by the inheritance of degenerate qualities; but sometimes it may be caused by environmental factors which are not hereditary. In the case given, the evidence points strongly toward the feeble-mindedness and moral delinquency of Carrie Buck being due, primarily, to inheritance and not to environment. (Laughlin, 1929, p. 17)

When I examined the records of Carrie Buck's case in the Amherst County courthouse, the direct impact of the Kallikak study on the trial was obvious. Dr. J.S. DeJarnette, the superintendent of Western State Hospital in Staunton, Virginia, testified on behalf of the state. When asked to give evidence that feeblemindedness was hereditary, he cited the Kallikaks. The transcript of his statement is made even more interesting by his incorrect recollection of the story and by the court reporter's problems with the spelling of the family name:

> An illustration was had in New Jersey, called the Call-icac case. Old man John Callicac in 1775 had an illegiti-mate child by a feeble-minded woman. He also had offspring from his wife, and none of them were feeble-minded. There were 480 offspring as a result of the child he had by this feeble-minded woman—(Dr. DeJarnette at this point consults his notes) there were 143 feeble-minded, 44 normal, and 293 undetermined; probably couldn't get the history on them—this occurred in 1775. That is a report that was generally published through-out most of the books on heredity. (Buck v. Priddy, 1924, p. 45)

DeJarnette was then questioned by the lawyer represent-ing the state colony concerning the Kallikaks:

> Question: In other words, the ancestor Callicac was normal?
> Answer: Supposed to be.

> Question: Mated with a wife that was normal?
> Answer: Yes, sir, and had 496 descendants.

> Question: None of them feeble-minded?
> Answer: No.

> Question: None criminal?
> Answer: Not so far as we know.

> Question: And then he mated with a feeble-minded girl?
> Answer: And their descendants were 480—143 of them were dependents of the state of New Jersey—that is about one-fourth—that is a little over one-fourth.

Question: Now on the side of the mating with the normal woman, what was the type of offspring as illustrated in the examples?
Answer: Normal all the way through.

Question: Did any of them reach eminence, do you
know?
Answer: I don't remember. (Buck v. Priddy, 1924,
pp. 46–47)

Arthur Estabrook not only collected the information on
Carrie Buck's family that was used in Laughlin's deposition,
he also testified at the trial. The state colony's counsel, A.E.
Stroud, came back to the Kallikaks when he questioned
Estabrook:

Question: Dr. DeJarnette made some references to the
Callicac family of New Jersey. Have you any knowl-
edge of the history of that family?
Answer: I have.

Question: I wish you would supplement what Dr.
DeJarnette said about them.
Answer: The only point to be added is that on the good
side of the Callicac family there were found among
the members several that have been college presi-
dents, at least one governor of the state, and a
number of senators.

Question: Were there any such types found among
those who were descended from the feeble-minded?
Answer: No. (Buck v. Priddy, 1924, p. 73)

The successes of the favored Kallikaks had increased since
1912! Estabrook's testimony (erroneous as it was) painted an
even more glorious picture of the effect of good genes than
Goddard's portrait had.

Thus, even though Goddard himself had been mild and
cautious on the question of compulsory sterilization, his

myth was very instrumental in legitimizing it. In the process of serving as an example of why sterilization laws were needed, the myth grew to even larger proportions.

In his deposition, Harry Laughlin also referred to Goddard's work. It is clear that the intent of his statement was to influence the court through the scientific credibility that Goddard's research gave to the arguments being made in favor of sterilization:

> I submit herewith Bulletin No. 1 of the Eugenics Record Office, by Henry H. Goddard, on the subject "Heredity of Feeble-mindedness." This bulletin of 14 pages contains 15 pedigree charts showing the family distribution of feeble-mindedness in these families, and demonstrating the hereditary nature of the defect. At the time of preparing this bulletin, namely in 1911, Dr. Goddard was the director of the scientific studies then being conducted in feeble-mindedness by the Training School at Vineland, N.J. (Buck v. Priddy, 1924, p. 11)

On page 13 of the cited bulletin is the pedigree chart of Deborah Kallikak. The chart shows only the disfavored Kallikak line; it was prepared before the connection to the favored line was discovered, or at least before it was revealed.

Attorney Stroud called several witnesses who knew some of Carrie's relatives. His purpose in having them testify was to demonstrate that defectiveness was spread throughout the family and was, indeed, hereditary in Carrie's case. These witnesses were obtained through investigations of the family conducted by Estabrook. As one reads the testimony, the techniques that Estabrook used and the manner in which he attempted to influence people's judgments about his fieldwork become obvious.

Three elementary school teachers testified during the trial. As each appeared before the court, a pattern developed in their remarks. The first to speak was Eula Wood. She was asked to talk about Carrie's younger half sister, Doris. She explained that she had been Doris's teacher for only 6 weeks and, therefore, had only limited knowledge of the child. Stroud asked if Miss Wood would call her a dull child. She replied, "Well, she is dull in her books—I would call her dull in her books." (Buck v. Priddy, 1924, p. 14)

Following Miss Wood, a second teacher, Miss Virginia Beard, was asked to give information about Roy Smith, Carrie's half brother. She testified that he did not do passing work in her fourth-grade class. She said that his behavior in school was a problem because he "tried to be funny—tried to be smart" (Buck v. Priddy, 1924, p. 15). When Stroud asked Miss Beard about how Roy would compare with other boys his age in school:

Answer: Well he is below the grade of other boys of his age in school.

Question: Basing your reply on your experience as a school teacher, would you consider him weak-minded?

Answer: Well, I don't know. (Buck v. Priddy, 1924, p. 16)

The third teacher, Miss Virginia Landis, was asked to give her assessment of the mental ability of one of Carrie's cousins. She said that she considered him "a dull child, but a normal child." (Buck v. Priddy, 1924, p. 17) When asked to explain, she focused on his poor school work saying that he was dull because he was slow in grasping his work in school and that he dropped out while in the fifth grade.

The impression I have from reading these court transcripts is that A.E. Stroud was seeking a clear pronouncement of feeblemindedness from these teachers and was far from satisfied with the qualified statements they gave him about the intellectual levels of Carrie's relatives. Stroud's examination of the next witness is indicative of his frustration at not getting the concrete and simple diagnosis he was seeking.

John W. Hopkins, a neighbor of Carrie's family and superintendent of the Albermarle County Home, was asked to talk about Carrie's brother, Roy Smith. He explained that the only thing he could say was that he had once observed the boy acting strangely on his way to school one day. This seems to have disturbed Stroud:

Question: Did you tell Dr. Estabrook that you consider that boy mentally defective and foolish?
Answer: I think so, yes.

Question: Then why don't you tell us that, then, Mr. Hopkins? Are you averse to testifying?
Answer: No, sir, but that is all I know about him. (Buck v. Priddy, 1924, p. 21)

Hopkins was then asked to testify about another relative on Carrie's mother's side of the family:

Question: What do you know about Richard Dudley?
Answer: Well I don't know very much about Mr. Dudley. He strikes me as being right peculiar, and that is all I know about him, but as to why, I couldn't tell you any particular case at all.

Question: Is he a man above, or below, the average intelligence?
Answer: Well I don't know sir. I don't know whether I am capable of judging that. . . .

Question: Didn't you tell Dr. Estabrook yesterday. . . .
Answer: I did—I told him I thought so, but since considering that thing. . . .

Question: It is natural that it would be embarrassing to you to testify about these people being neighbors—
Answer: I know, but I don't mind telling you what I know to be fact. (p. 21)

In asking Hopkins to testify about the relative's son, Stroud uses the same type of questioning:

> Question: Do you consider him above or below the average?
> Answer: Well, that question is exactly like the other, and I answer it the same way.
>
> Question: Yesterday you thought he was below, and today you don't know?
> Answer: Well, I don't know. That is right. (pp. 21–22)

Samuel Dudley was another neighbor called to the trial. As it turned out, apparently to Stroud's and Estabrook's surprise, he was also Carrie's great uncle. He was asked to make comments on Carrie's grandfather, Richard Harlow:

> Question: What did you—what was your opinion of Richard mentally?
> Answer: I suppose Richard had just as good ordinary sense as the generality of the people. Now, Mr. Stroud, he wasn't a thorough educated man. He had some little joking ways sometimes, but outside of that he was all right.
>
> Question: Did you regard him as at all peculiar in any way?
> Answer: No, no more than just in a joking manner, sir.
>
> Question: Didn't you tell Dr. Estabrook yesterday or the day before, that you considered Richard peculiar, or below the average?
> Answer: No, sir, I just told him that he had those peculiar ways. That gentleman there (pointing) asked me Saturday night, and pressed me about a lot of things I didn't know anything about.
>
> Question: Didn't you tell him you thought Richard was peculiar or below the average?

Answer: Just in this joking way and the manner he had. He was a man that transacted his own business up until his death.

Question: But you did tell Dr. Estabrook he was peculiar?

Answer: Well possibly I did. He kept quizzing me about different things, and I thought I would just let him go. (Buck v. Priddy, 1924, pp. 25–26)

The testimony of Miss Caroline Wilhelm, a social worker, probably comes closest to telling the truth of Carrie Buck's case. It is unlikely that Carrie would have been institutionalized if she had not gotten pregnant and had an illegitimate child. The combination of her poverty, lack of a protecting family group, limited education and skills, her youth, and her pregnancy resulted in her commitment to the state colony. Her foster family initiated the procedure to institutionalize her only when they discovered that she was pregnant. Had there been some means for her to hide, explain, or legitimize her condition, Carrie Buck would not have been classified as feebleminded. Nor would she have become the victim of and precedent for compulsory sterilization. Miss Wilhelm's statements demonstrate the circular and inescapable reasoning that led to Carrie's separation from society and her eventual sterilization:

Question: Now, there are records down in Charlottesville in connection with social work—have they any records against Carrie Buck, the girl here, which would tend to show that she was feeble-minded or unsocial or anti-social, or whatever the term is, other than the birth of this child?

Answer: No sir, our record begins on the 17th of January of this year, and that is the first knowledge we have of her.

Question: Basing your opinion that the girl is unsocial or anti-social, on the fact that she had an illegitimate child—the point I am getting at is this—are you basing your opinion on that?
Answer: On that fact, and that as a social worker I know that girls of that type—

Question: Now, what is the type?
Answer: I should say decidedly feeble-minded.

Question: But the question of pregnancy is not evidence of feeble-mindedness, is it—the fact that, as we say, she made a miss-step—went wrong—is that evidence of feeble-mindedness?
Answer: No, but a feeble-minded girl is much more likely to go wrong. (Buck v. Priddy, 1924, pp. 33–34)

Carrie Buck's trial was obviously a test case arranged by the administration of the state colony, probably with the encouragement and support of the Eugenics Record Office. The man who served as Carrie's guardian in the case, R.G. Shelton, was appointed at the request of the colony. No witnesses were called on Carrie's behalf. The only arguments in her favor were references to legal questions, and these were brief and technical. But perhaps the most telling part of the case was a comment from A.S. Priddy, superintendent of the state colony.

Priddy gave examples of people at the colony who had been sterilized for medical reasons and who had been placed successfully outside the institution. In an aside, he indicated that Carrie's attorney, Mr. Whitehead, knew the "inmates" he was talking about. Mr. Whitehead acknowledged quickly that he did know these people. He seems to have wanted to clarify why he knew them; he told the court reporter, "Yes, put in there that I know them through being a member of the Special Board of Directors (Buck v. Priddy, 1924, p. 90). Carrie Buck's counsel, then, was not a real counselor for her but was, instead, merely another player in a legal charade.

Attorney Stroud reported to the court that he had interviewed Carrie concerning sterilization. Years later, Carrie would say that she had not understood at the time what the operation would do to her. But Stroud, in reporting on his conversation with her, seemed to want to convey the impression that Carrie was unconcerned about being sterilized:

> Question: Do you care to say anything about having this operation performed on you?
> Answer: No, sir, I have not, it is up to my people. (Buck v. Priddy, 1924, p. 29)

Carrie must have thought that her family would be involved in making the decision of what was best for her. Tragically, in this case, Carrie had no "people."

The court decided in favor of the state and ordered Carrie's sterilization. The case was appealed in Virginia, and the circuit court's decision was upheld. Finally, the case was heard by the U.S. Supreme Court. The majority of the court ruled that Virginia's compulsory sterilization law was constitutional (Buck v. Bell, 1927). Thus, the precedent was set giving state governments the right to intervene in the reproductive practices of citizens who were deemed defective in some way.

In delivering the majority opinion in this landmark decision, Justice Oliver Wendell Holmes said:

> We have seen more than once that the public welfare may call upon the best citizens for their lives. It would be strange if it could not call upon those who already sap the strength of the State for these lesser sacrifices, often felt to be much by those concerned, in order to prevent our being swamped with incompetence. It is better for all the world, if instead of waiting to execute degenerate offspring for crime, or to let them starve for their

imbecility, society can prevent those who are manifestly unfit from continuing their kind. The principle that sustains compulsory vaccination is broad enough to cover cutting the Fallopian tubes. . . . Three generations of imbeciles are enough. (Buck v. Bell, 1927, p. 50)

During the 1970s, Dr. K. Ray Nelson was director of the institution in which Carrie Buck was sterilized. It is now known as the Central Virginia Training Center. While administering the programs at the institution, he began to do research on the center's records relating to the practice of sterilization. He was interested in finding out what he could about Carrie and her sister Doris, who had also lived in the institution and been sterilized there. In the process of reviewing the files, he discovered that, following the Buck v. Bell decision and Carrie's sterilization, more than 4,000 people had been sterilized in that one institution. The practice was not completely abandoned there until 1972.

Nelson finally found Doris and Carrie in 1979. Doris Buck Figgins and her husband, Matthew, were living near Front Royal, Virginia. They had been married for 39 years. In his visit, Nelson brought copies of institutional records in which Doris was interested. He read to her some important dates, including her birthdate, which she had not known. When he read the date of her sterilization, he heard a cry; looking up from the records he was reading, he found Doris and Matthew sobbing:

"They didn't know she'd been sterilized," Dr. Nelson said. Mrs. Figgins told him that when she was wheeled into an operating room at the Lynchburg hospital at the age of 16 in 1928, doctors indicated only that they were going to perform an appendectomy.

"Here was this lady," Dr. Nelson said, "who for years had been feeling that she had failed because she couldn't have children." (Robertson, 1980, pp. A-1, D-4)

Through Mrs. Figgins, Nelson was able to locate Carrie Buck Detamore, who was living at that time in Albemarle County near Charlottesville, Virginia. In subsequent newspaper interviews, Carrie told reporters that the operation was performed on her because she got pregnant and had a baby by a boy friend who had raped her. She explained that she was told only that she had to have an operation, not that it would mean that she would never be able to have children again. She did not find out that she had been sterilized until several years after the operation was performed. "I didn't want a big family," she said in an interview, but "I'd like to have a couple of children." She said, "I was surprised. Oh yeah, I was angry. . . . I just didn't like the idea of being operated on to keep from having children" ("Case Led," 1980, p. 24).

Paul A. Lombardo of the University of Virginia, a legal scholar, has studied the Buck v. Bell case. He visited Carrie Buck during the last years of her life. In a letter to Stephen Jay Gould, he remarked:

> As for Carrie, when I met her she was reading newspapers daily and joining a more literate friend to assist at regular bouts with the crossword puzzles. She was not a sophisticated woman, and lacked social graces, but mental health professionals who examined her in later life confirmed my impressions that she was neither mentally ill nor retarded. (Gould, 1984, p. 16)

The statement by Harry Laughlin recorded in the original circuit court's proceedings includes only a brief mention of Carrie Buck's daughter (the "third generation of imbeciles"), who became so important eventually in reaching the Holmes decision. Carrie was said to have "had one illegitimate child, now about six months old and supposed to be mental defective" (Laughlin, 1929, p. 16).

Carrie's baby was adopted by the same people who had raised her—J.T. and Alice Dobbs. It was the comparison of this baby with Dobbs's own grandchild on which the diagnosis of mental deficiency was based. Miss Wilhelm, the social worker, testified in the circuit court trial that the baby did not seem quite normal to her. At the time of the original trial, the baby was only 7 months old. Miss Wilhelm said that, compared to the Dobbs grandchild, who was only a few days older than Carrie's baby, there was a decided difference in development: "There is a look about it that is not quite normal, but just what it is, I can't tell" (Gould, 1984, p. 17).

Little has been known about Carrie's child until the recent collaborative efforts of Stephen Jay Gould and Paul Lombardo. In a startlingly revealing article, Gould explains that Carrie's child was a girl named Vivian who continued to live with the Dobbs family throughout her short life. She died at the age of 8 from a not clearly diagnosed childhood disease. Before her death, Vivian attended public school for four terms, from September 1930 until May 1932. The records from her school indicate that she was a normal little girl and an average student. She progressed well in her academic subjects and consistently received high marks for "deportment." In the spring of 1931, she was on the honor roll of her school. Both Gould and Lombardo conclude that, not only was Vivian a quite normal child, but there is no reasonable evidence that either Carrie or her mother was retarded. In other words, the existence of "three generations of imbeciles" was patently untrue.

Gould concludes his revisit to the Carrie Buck case with words that are well worth repeating here:

> Carrie Buck died last year. By a quirk of fate, and not by memory or design, she was buried just a few steps from her only daughter's grave. In the umpteenth and ultimate verse of a favorite old ballad, a rose and a brier—the sweet and the bitter—emerge from the tombs of Barbara Allen and her lover, twining about each other in the union of death. May Carrie and Vivian, victims in

different ways and in the flower of youth, rest together in peace. (Gould, 1984, p. 18)

Following Nelson's disclosure of the number of steriliza-tions that had been performed at the Lynchburg institution, a series of newspaper articles appeared on the subject. One included an interview with a man who had lived there and who had been sterilized when he was 15 years old. In the article he is called Buck Smith.

Buck was born in Richmond to a couple who could not support him. He lived his first 8 years in various city institu-tions. For reasons he does not understand, he was then sent to the Lynchburg colony:

"They separated us according to ability," Smith recalled. "Most of the kids seemed to come from broken homes. There wasn't, to my mind, that many retarded. They were just sort of lost. . . .

"Eventually, you knew your time would come," he said. "Everybody knew it. A lot of us just joked about it. There was a lot of kidding and joking. We weren't growed up enough to think about it. We didn't know what it meant." (McKelway, 1980, pp. A-1 & B-1)

Buck recalled that one day a group of boys and girls met in the basement of one of the buildings:

"We were just beginning to find out what life was all about," Smith said, explaining that some sexual activity took place.

"Then the girls told on us," he said, "and they put me in confinement for a month. They said I had to be taught a lesson. Two weeks later they came to me and told me they were going to have to sterilize me." (McKelway, 1980, p. B-1)

Buck goes on to describe his recollections of the steriliza-tion procedure and of his departure from the institution a short time later. He was married at age 18. That marriage ended in divorce after 13 years. He felt that his wife was unable to accept the fact that they could not have children, and yet could not bring herself to talk with him about her feelings. Buck remarried shortly afterward to a woman who had two children by a previous husband. The article closes with Buck Smith's expression of regret at not having children of his own:

> "Having children is supposed to be part of the human race. Sometimes I feel like there's a part that I'm missing."
>
> Tears welled up in his eyes, surrounded by creases of a life of hard work.
>
> Behind him, pasted to a mirror, was a dollar bill Smith said might bring him luck. And below that was a card from his stepchildren.
>
> "Thinking of you Daddy," it reads. "They call me Daddy," Smith said. (McKelway, 1980, p. B-1)

As this chapter is being written in July of 1984, a proposed settlement in a class action suit has just been filed in the U.S. District Court in Lynchburg, Virginia, on behalf of people who were involuntarily sterilized in Virginia. The terms of the settlement provide for a media campaign to notify former residnts of state institutions that they can inquire and determine if they were sterilized. They also provide for psychological counseling for persons who were sterilized against their will or without their knowledge—seemingly small restitution for a great injustice. But even greater injustices must be examined in order to understand the legacy of the Kallikaks.

In 1927, the model sterilization act developed by Harry Laughlin and used by Virginia was held to be constitutional by the U.S. Supreme Court. On July 14, 1933, the same model became law in Germany. On that day Adolf Hitler decreed that the Hereditary Health Law was in force. The law was intended to ensure that "less worthy" members of the Third Reich did not pass on their inferior genes. Hereditary health courts were established to decide which persons were to be sterilized. Each court was to consist of two doctors and one judge—all government appointed (Ludmerer, 1972).

The Law for the Protection of German Blood and Honor
September 15, 1935

Imbued with the knowledge that the purity of German blood is the necessary prerequisite for the existence of the German nation, and inspired by an inflexible will to maintain the existence of the German nation for all future times, the *Reichstag* has unanimously adopted the following law, which is now enacted:

Article 1. (1) Any marriages between Jews and citizens of German or kindred blood are herewith forbidden. Marriages entered into despite this law are invalid, even if they are arranged abroad as a means of circumventing this law.

   (2) Annulment proceedings for marriages may be initiated only by the Public Prosecutor.

Article 2. Extramarital relations between Jews and citizens of German or kindred blood are herewith forbidden.

Article 3. Jews are forbidden to employ as servants in their households female subjects of German or kindred blood who are under the age of forty-five years.

Article 4. (1) Jews are prohibited from displaying the Reich and national flag and from showing the national colors.

   (2) However, they may display the Jewish colors. The exercise of this right is under state protection.

Article 5. (1) Anyone who acts contrary to the prohibition noted in Article 1 renders himself liable to penal servitude.

   (2) The man who acts contrary to the prohibition of Article 2 will be punished by sentence to either a jail or penitentiary.

   (3) Anyone who acts contrary to the provisions of Articles 3 and 4 will be punished with a jail sentence up to a year and with a fine, or with one of these penalties.

Article 6. The Reich Minister of Interior, in conjunction with the Deputy of the *Fuehrer* and the Reich Minister of Justice, will issue the required legal and administrative decrees for the implementation and amplification of this law.

Article 7. This law shall go into effect on the day following its promulgation, with the exception of Article 3, which shall go into effect on January 1, 1936.

The German law was implemented swiftly and broadly. By the end of the first year that the law was in effect, over 56,000 people had been found to be defective by the health courts and were sterilized (Holmes, 1936). Hitler's actions were applauded by American eugenics proponents. Paul Popenoe felt that the Germans were following a policy that was consistent with the thinking of eugenicists throughout the world (Popenoe, 1934). Ludmerer quotes an editorial statement from the *Eugenical News* that concluded, "It is difficult to see how the new German Sterilization Law could, as some have suggested, be deflected from its purely eugenical purpose, and be made an 'instrument of tyranny' for the sterilization of non-Nordic races" (1972, p. 117).

It has been estimated that, between 1933 and 1945, two million people were deemed defective and sterilized in Germany. In testimony at the Nuremberg war trials, a German doctor cited Virginia's Buck vs. Bell case as the precedent for Nazi race hygiene and sterilization programs (Booker, 1980).

In 1935, the Nazi government passed the Nuremberg Laws. These were based on the continuing German research in *Rassenhygiene* (race hygiene). The laws banned interracial marriage between Germans and Jews and elaborated on the original sterilization act. The articles of the laws addressing the issue of interracial marriage are chilling in the extent to which they reflect the influence of the American eugenics movement (Snyder, 1981, pp. 213–214):

If, at this point, the connection between the eugenics movement in America and the Nuremberg laws is not yet clearly traceable, it may be helpful to consider the *Act to Preserve Racial Integrity*, which was enacted in Virginia in 1924. The act was written and guided through the state legislature by W.A. Plecker. Plecker, a strong believer in eugenics, served for many years as the registrar of vital statistics for Virginia. He worked closely with the Eugenics Record Office and was a member of several eugenics organizations. A.H. Estabrook called upon Plecker for assistance in a study of racially mixed families. In his book, *Eugenics in*

Relation to the New Family, Plecker quotes Estabrook in language typical of the eugenicists:

> Dr. A.H. Estabrook in a recent study for the Carnegie Institute, of a mixed group in Virginia, many of whom are so slightly negroid as to be able to pass for white, says: "School studies and observations of some adults indicate the group as a whole to be of poor mentality, much below the average . . . on the basis of the army intelligence tests. There is an early adolescence with low moral code, high incidence of licentiousness and twenty-one percent of illegitimacy in the group." (Plecker, 1924, p. 15)

The Virginia *Act to Preserve Racial Integrity* states in part:

> It shall hereafter be unlawful for any white person in this State to marry any save a white person, or a person with no other admixture of blood than white and American Indian. For the purpose of this act, the term "white person" shall apply only to the person who has no trace whatsoever of any blood other than Caucasian; but persons who have one-sixteenth or less of the blood of the American Indian and have no other non-Caucasian blood should be deemed to be white persons. All laws heretofore passed and now in effect regarding the inter-marriage of white and colored persons shall apply to marriages prohibited by this act. (Plecker, 1924, p. 31)

Shortly after the Nazis took power, they took control of all major German universities. In June 1936, Heidelberg University held a celebration commemorating its 550th anniversary. Honorary degrees were awarded to a number of European and American scholars. Harry Laughlin was one of those honored. The degree was conferred in appreciation of his services to the science of eugenics and his efforts to purify "the human seed stock." Laughlin's invitation from the dean of the Heidelberg faculty of medicine reads:

The Faculty of Medicine of the University of Heidelberg intends to confer upon you the degree of Doctor of Medicine h.c. (honoris causa) on the occasion of the 550 year Jubilee (27th to 30th of June, 1936). I should be grateful to you if you would inform me whether you are ready to accept the honorary doctor's degree and, if so, whether you would be able to come to Heidelberg to attend the ceremony of honorary promotion and to personally receive your diploma. (Schneider, 1936)

Laughlin responded with dispatch and enthusiasm:

I stand ready to accept this very high honor. Its bestowal will give me particular gratification, coming as it will from a university deep rooted in the life history of the German people. . . . To me this honor will be doubly valued because it will come from a nation which for many centuries nurtured the human seed-stock which later founded my own country and thus gave basic character to our present lives and institutions. (Laughlin, 1936a)

After the degree was awarded, Laughlin again wrote to the dean expressing his deep appreciation for the honor:

I consider the conferring of this high degree upon me not only as a personal honor, but also as evidence of a common understanding of German and American scientists of the nature of eugenics as research in and the practical application of those fundamental biological and social principles which determine the racial endowments and the racial health—physical, mental and spiritual—of future generations. (Laughlin, 1936b) [1]

Marian S. Olden of the Association for Voluntary Sterilization recalls in positive terms her exposure to the Nazi sterilization program:

At home I had read everything available on the subject
and had a well founded conviction that it was administered scientifically and rationally, not emotionally or
racially.

[The law] had been extensively discussed and
approved by the Prussian Ministry of Public Health and
Social Welfare in 1932. It stipulated that every doctor in
the nation must report every case of hereditary disease
that came to his attention. The law applied to: 1) Con-
genital Feeble-mindedness, 2) Schizophrenia,
3) Manic-depressive insanity, 4) Inherited epilepsy,
5) Huntington's chorea, 6) Severe hereditary malfunc-
tions, 7) Severe alcoholism, 8) Hereditary blindness,
9) Hereditary deafness. . . .

A good sterilization law must carry safeguards for its
proper administration. . . . If sterilization were done
without medical or eugenic indications, it was considered malpractice and would be prosecuted under the
German Criminal Code. In Germany in 1937 there were
196 Health Courts, functioning quite apart from the
Criminal Courts. Each Health Court was composed of a
district judge, a public health officer, and a physician
specializing in medical genetics. They conducted in-
quiries into the condition of the entire family as well as
examinees the person brought before them. (Olden,
1974, p. 65)

It is important to note that Ms. Olden related these positive
impressions of the Nazi program in a book published in 1974.
The book is essentially a contemporary version of the original eugenic arguments for sterilization. Indeed, eugenics is
still flourishing in many ways, as we document in the following chapter. Even the realities of the Nazi race hygiene
program that were revealed following World War II did not
shake Olden's faith in eugenics and sterilization. In her
book, she speaks with great pleasure of her meeting with
Henry Goddard, "the author of the famous little book *The*

Kallikak Family which in the 20's did much to promote interest in the subject of eugenics" (Olden, 1974, p. 95).

Olden includes quotations from Dr. Marie Kopp, an American of Swiss heritage who received a fellowship to study the administration of the German sterilization program in 1935. Kopp reportedly traveled widely in Germany and interviewed a wide assortment of people. According to Olden, Kopp reported:

> The sterilization law is accepted as beneficial legislation, designated to minimize the difficulties of the afflicted. All possible safeguards are taken to forestall miscarriages of justice in whatever form they may occur. . . . I am convinced that the law is administered in entire fairness and that discrimination of class, race, creed, political or religious belief does not enter into the matter. I say this with confidence. (Olden, 1974, p. 65)

While looking through the papers of Edgar Doll in the Archives of the History of American Psychology, I discovered an interesting aside to Marie Kopp's enthusiastic appraisal of the Nazi sterilization program. Kopp wrote to Doll, Goddard's assistant and eventual successor at the Vineland Training School, asking for the real name of the Kallikaks and any other information he could provide about the family. Like me, she was interested in revisiting the Kallikaks. I imagine, however, that the nature of her interest was quite different from mine. In his reply, Doll did not address her request for the real name. He referred her to Elizabeth Kite as a possible source of information. He also commented that "we have done nothing on this since the book was published except for a somewhat casual field followup by Dr. Goddard in 1917" (Doll, 1941, p. 134).

With respect to the strongly supportive views of Olden and Kopp regarding the Nazi sterilization program, we know that in fact the program was administered in a racist and capricious fashion. People from groups held in disfavor by the government were much more likely than others to be found to be defective in some way and sterilized. Anyone who was not Nordic in background and appearance was at risk. Allan Chase cites the observations of Wallace Duell, *Chicago Daily News* correspondent:

> In examining supposedly feeble-minded persons to decide whether or not they are subject to compulsory sterilization, the Nazis gave them an intelligence test devised by the Reich government. The tests, however . . . had to be frequently changed since the supposedly feeble-minded persons who took them were passing answers along to their friends. This did them little good, as they would now be sterilized for moral deficiencies. As the official report enforcing the sterilization law (quoted by Duell) indicated: "Among the feeble-minded there is a large number who have a certain mental agility and who answer the usual easy questions quickly and apparently with assurance, and who only after a more searching examination betray the utter superficiality of their thinking and their inability to reason and their lack of moral judgment." (Chase, 1977, p. 350)

Goddard's conception was enduring. Although it was clearly difficult at times, some way could always be found to create a moron.

Die Familie Kallikak, the first German edition of Goddard's book, was published in 1914. The second edition was published in November 1933, after the Nazis came to power. The full text of this second edition, together with an introduction, appeared in a special issue of an academic journal, *Friedrich Mann's Pedagogisches Magazin.* The translator, Karl Wilker, makes very clear the impact of the Kallikak study in Germany:

> The first printing of this book aroused considerable attention. This attention often was expressed even in

doubts about the genuineness of the study. How could the history of this family be true when they lived in the "land of unlimited possibilities." In the meantime research on genetic inheritance has undergone an entirely unforeseen blooming. Questions which were only cautiously touched upon by Henry Herbert Goddard at that time . . . have resulted in the law for the prevention of sick or ill offspring dated the 14th of July, 1933 [the sterilization law]. These questions then, have since become generally interesting and significant. Just how significant the problem of genetic inheritance is, perhaps no example shows so clearly as the example of the Kallikak family. (Wilker, 1934, p. ii)

At the end of the book there is an illustration portraying the two Kallikak lines. The illustration was apparently done in 1925 by Professor Martin Fogel of the German Hygienic Museum. Fogel's rather Gothic-looking representation, reproduced here as Illustration 13, is interesting in the manner in which the "good" and "bad" Kallikaks are pictured: The favored line has the appearance of a robust Nordic group; the disfavored line seems to look less Aryan.[2]

In *Justice at Nuremberg*, Robert Canot quotes the observation of a German medical economist:

The care of a deaf-mute or cripple costs 6 marks a day, that of a reform school inmate 4.85 marks, and that of a mentally ill or deficient person 4.5 marks. The average earnings of a laborer, on the other hand, were only 2.5 marks, and those of a civil servant 4 marks daily. . . . The state spends far more for the existence of these actually worthless compatriots than for the salary of a healthy man, who must bring up a healthy family. (Conot, 1983, p. 205)

The compulsory sterilization laws of Nazi Germany set the stage for what was to become the most comprehensive and vigorous eugenics program the world had ever known. In

Illustration 13: Pictorial representation of two lines of the Kallikak family. Reprinted with permission from "Die Familie Kallikak," trans. K. Wilker, 1934, *Friedrich Mann's Pedagogisches Magazin,* No. 1393.

1937, Hitler authorized certain officially appointed physicians to grant mercy deaths to incurable persons. The first to be given this grant were physically handicapped children in a hospital near Wurthemberg. The children were killed with overdoses of drugs mixed in their food. Those who would not eat were killed with injections or suppositories. Soon, questionnaires were sent to all institutions that housed children. On the basis of the questionnaire results, children who were deemed incurable or genetically defective were picked for "besondere Heilverfahren" (special healing procedures). In 1940, the exterminations were expanded to include handicapped adults. Every institution caring for mentally, emotionally, or physically disabled persons was required to fill out patient questionnaires. Doctors and medical students selected the people they judged to be incurable or genetically tainted, and those selected were then transported to a euthanasia center. Soon the selection procedure became a mere formality; the institutions were cleared en masse.

According to a physician who worked in the program:

Most institutions did not have enough physicians, and what physicians there were were either too busy or did not care, and they delegated the selection to the nurses and attendants. Whoever looked sick or was otherwise a problem patient from the nurses' or attendants' point of view was put on a list and was transported to the killing center. (Conot, 1983, p. 207)

At the beginning of 1941, William Shirer, then a correspondent for *Life,* reported:

Never related before in this country and known to few people in Germany itself, has been the execution of tens of thousands of the mentally deficient throughout the Reich. Few details of these fantastic "mercy killings" are known, but it has been established that the Gestapo is now carrying out the systematic murder of thousands of mental misfits dragged from both private and state sanitariums. Only Hitler and a few men at the top—and of course the relatives who are told to fetch the ashes—know of it yet. (Hackett, 1941, p. 138)

The mass elimination of handicapped people—that is, of people judged to be defective in some way—was not a unique Nazi invention. It was the culmination of a eugenic philosophy that had been building in strength for decades. As one observer has pointed out, it was not a specific German creation or even a Nazi creation but a phenomenon of western thought and science (Wolfensberger, 1981).

Years before Hitler came to power, a euthanasia program was proposed by two German professors in a book titled *The Release of the Destruction of Life Devoid of Value*. Writing in the eugenic tradition, the authors point to the economic drain created in society by defective people and to the social costs such people extract from fit citizens. Their work developed the perception of the weak, poor, and handicapped as "useless eaters" and "superfluous people." The message was that there should be a social obligation to find and eliminate the misfits and the unfit (Binding & Hoche, 1975).

The Nazi eugenics program was explained to the young people of Germany in a government publication, the *Official Handbook for Schooling the Hitler Youth*. Seven million copies of the book were distributed to youngsters between 14 and 18 years of age. A chapter on heredity stresses the central role of inheritance in the perpetuation of desirable and undesirable human traits. It also gives an extensive list of defects that are "known" to be hereditary (Hackett, 1941). Noting that there were one million feebleminded people in Germany at that time, the handbook discusses the implications of the increasing number of defectives:

Most of these congenitally diseased and less worthy persons are completely unsuited for living. . . . They cannot take care of themselves and must be maintained and cared for in institutions. This costs the state enormous sums yearly. And in this connection some figures might well be given. The outlay for an inmate of an institution for hereditary diseases is eight times as high as it is for a sound person. Just about as much money is needed for an idiotic child as for four or five sound children. The instruction of a pupil for eight years costs about 1,000 marks, the educational outlay for a deaf mute about 20,000 marks. Altogether Germany pays every year about 1,200,000,000 marks to care for and support comrades afflicted with hereditary maladies. (Hackett, 1941, pp. 140–141)

On the basis of economics and the taint of bad blood, then, the necessity of killing handicapped people was justified. The same arguments would shortly be used as rationales for the elimination of Jews. In fact, most authorities on the Nazi program of genocide point to the fact that the killing of handicapped people evolved into the devastation of Europe's Jews (Wolfensberger, 1981). The philosophy, personnel, and equipment—and the deadened consciences—required for the Holocaust were developed through the process of killing handicapped people, those who were perceived to be defective, and those who were assumed to be morons.

It has been estimated that as many as a million people died in the Nazi euthanasia campaign against "defectives." About 400,000 of these were classified as either mentally ill or mentally retarded (Wertham, 1966). The program was very comprehensive and effective. Indeed, the program was so successful that Wolfensberger (1981) notes:

my visit to a large German institution for the mentally retarded in 1963 revealed the presence of relatively few living units for mature adults because few mentally retarded adults were then to be found in Germany (p. 3).

NOTES

1. The correspondence between Harry Laughlin and Carl Schneider quoted here first appeared in a report by Randall Bird and Garland Allen in the *Journal of the History of Biology* (Fall 1981), Vol. 14, No. 2.
2. I would like to express my appreciation to Professor Wayne Thompson, formerly of Lynchburg College in Virginia, now at the Virginia Military Institute, for his assistance in translating material from *Die Familie Kallikak.*

REFERENCES

Binding, K. & Hoche, A. (1975). *The release of the destruction of life devoid of value.* R. Sassone, Ed. Santa Ana, Calif.: Life Quality Paperbacks. Originally published, 1920.

Booker, B. (1981, February 27). Nazi sterilizations had their roots in U.S. eugenics. *Richmond Times-Dispatch,* Richmond, Va., pp. A 1, A 6.

Buck v. Bell, 274 U.S. 200, 47 S.Ct. 584 (1927).

Buck v. Priddy. (1924). Amherst County Clerk of Courts Office, Amherst County Courthouse, Amherst, Virginia.

Case led to sterilization law. (1980, February 27). *The Daily Advance,* Lynchburg, Va., p. 24.

Chase, A. (1977). *The legacy of Malthus: The new scientific racism.* New York: Knopf.

Conot, R.E. (1983). *Justice at Nuremberg.* New York: Harper & Row.

Doll, E. (1941). [Letter]. *Doll papers* (Box M236, Correspondence with Marie Kopp). Akron, Ohio: University of Akron, Bierce Library, Archives of the History of American Psychology.

Gould, S.J. (1984). Carrie Buck's daughter. *Natural History,* 93, 7, 14–18.

Hackett, F. (1941). *What Mein Kampf means to America.* New York: Reynal & Hitchcock.

Holmes, S.J. (1936). *Human genetics and its social import.* New York: McGraw-Hill.

Laughlin, H.H. (1922). *Eugenical sterilization in the United States.* Chicago: Psychopathic Laboratory of the Municipal Court of Chicago.

Laughlin, H.H. (1929). *The legal status of eugenical sterilization.* Chicago: Psychopathic Laboratory of the Municipal Court of Chicago.

Laughlin, H.H. (1936a). [May 28 letter to C. Schneider]. *Laughlin Notebooks, Vol. II, Nazi Eugenics.* Washington University, St. Louis, Mo.

Laughlin, H.H. (1936b). [August 11 letter to C. Schneider]. *Laughlin Notebooks, Vol. II, Nazi Eugenics.* Washington University, St. Louis, Mo.

Ludmerer, K.M. (1972). *Genetics and American society.* Baltimore: Johns Hopkins University Press.

Marks, R. (1981). *The idea of I.Q.* Washington, D.C.: University Press of America.

McKelway, B. (1980, February 24). They gave me what life I have and they took a lot of my life away. *Roanoke Times & World News,* Roanoke, Va., pp. A-1, B-1.

Olden, M. (1974). *History of the development of the first national organization for sterilization.* Gwynedd, Pa.: Foulkeways.

Plecker, W.A. (1924). *Eugenics in relation to the new family.* Richmond: Virginia State Board of Health.

Popenoe, P. (1934). The German sterilization law. *Journal of Heredity, 25,* 257–264.

Popenoe, P., & Johnson, R. (1918). *Applied eugenics.* New York: Macmillan.

Robertson, G. (1980, February 24). Interest in patient rights led to sterilization data. *Richmond Times-Dispatch,* Richmond, Va., pp. A-1, D-4.

Schneider, C. (1936). [May 16 letter to H. Laughlin]. *Laughlin Notebooks, Vol. II, Nazi Eugenics.* Washington University, St. Louis, Mo.

Snyder, L. (1981). *Hitler's Third Reich: A documentary history.* Chicago: Nelson-Hull.

Voorhees, D.W. (Ed.). (1981). *Dictionary of American biography.* New York: Charles Scribners Sons.

Wallin, J.E. (1917). *Problems of subnormality.* New York: World Book Co.

Wertham, F. (1966). *A sign for Cain: An exploration of human violence.* New York: Macmillan.

Wilker, K. (Trans.). (1934). Die Familie Kallikak. *Friedrich Mann's Pedagogisches Magazin* (No. 1393) [Special Issue]. (Original work published 1914, 2nd ed. 1933).

Wolfensberger, W. (1981). The extermination of handicapped people in World War II Germany. *Mental Retardation, 19,* 1–7.

Chapter 10

The New Eugenics

AFTER WORLD WAR II, THE EUGENICS MOVEMENT DECLINED in popularity and power. The decline was due in large part to the growing awareness in scientific and intellectual circles of the awful realities of the Holocaust. The role of eugenic philosophy and practices in Nazi Germany alarmed those who saw the connection and led to much greater caution in the polemics surrounding the nature-nurture question. Even those who continued to believe in the tenets of eugenics were more careful and quiet in their advocacy of genetic solutions to social problems. The eugenics movement and the legacy of the Kallikaks were, however, far from dead.

Even in the awful light of the Nazi atrocities, the story of the Kallikaks persisted as a sort of primal myth. Henry Garrett, former president of the American Psychological Association and chairman of the Department of Psychology at Columbia University, continued to present the Kallikak study as a legitimate source of data supporting the genetic basis of intelligence. In his textbook, *General Psychology,* the Kallikak family is used to portray the connection between mental retardation, immorality, and genetics. Garrett included a drawing of Martin Kallikak's two lines of descendants: Those in the "bad" line are shown with horns and evil grins; those in the "good" line, as might be expected, wear Puritan hats and pious expressions (see Illustration 14).

Garrett's book was widely used as an introductory text in psychology courses. It is hard to imagine, and of course impossible to measure, the influence that the Kallikak myth continued to have on college students because of its authoritative legitimization by Garrett. There can be no doubt, however, that Garrett's textbook portrayal helped to preserve and foster the concept of hereditary defectiveness and degeneracy in a new generation of students—students who, in their turn, were to become leaders and shapers of future public opinion (Chase, 1977).

Thus, during the 1960s, Garrett carried the academic banner for eugenics. What might otherwise have been rejected as a simple prejudicial notion was given credence in many sectors because of Garrett's background and supposed expertise. Following his textbook portrayal of the Kallikaks, he wrote a series of pamphlets in which he described the degenerative effects that racial integration would have in the United States. In one of these, titled *Breeding Down,* he used two arguments that had long been standards of the eugenics tradition: one, that intelligence is primarily hereditary; second, that some groups have less of it than others. Stephen Chorover, in his important book, *From Genesis to Genocide,* quotes from the pamphlet:

Illustration 14: A textbook portrayal of the Kallikaks. Reprinted with permission from *General Psychology* (rev. ed., p. 65) by Henry Garrett, 1955, New York: American Book Company.

You can no more mix the two races and maintain the standards of White Civilization than you can add 80 (the average IQ of Negroes) and 100 (the average IQ of Whites), divide by two and get 100. What you would get would be a race of 90's, and it is that 10 percent differential that spells the difference between a spire and a mud hut; 10 percent—or less—is the margin of civilization's "profit"; it is the difference between a cultured society and savagery. Therefore, it follows, if miscegenation would be bad for White people, it would be bad for Negroes as well. For, if leadership is destroyed, all is destroyed. (Chorover, 1979, p. 47)

Desegregation of the schools must be prevented, argued Garrett, because it may encourage intermarriage, and intermarriage will destroy the purity and quality of the white race. The white race will be "bred down." Garrett's pamphlets were distributed free by opponents of integration to educators all over the country. In 1975, in the heat of the busing controversy, Henry Garrett's *IQ and Racial Differences* was published. An advertisement encouraged people to buy the book to gain "sufficient ammunition to answer and demolish . . . arguments for school integration point by point" (Chorover, 1979, p. 48).

In accepting the validity of Goddard's work, Garrett was not alone. In 1965, Sheldon and Elizabeth Reed of the University of Minnesota produced a voluminous work, *Mental Retardation: A Family Study*. The book contained information on over 80,000 people, all of them descendants of a group of 289 residents of the Faribault State School and Colony in Minnesota who had first been studied in 1911. The original study had been conducted under the auspices of the Eugenics Record Office. In their book, they wrote of Charles Davenport:

Dr. Charles B. Davenport had one of the most brilliant minds of the early day geneticists. It should be recalled that Mendel's laws were rediscovered in 1900. Many famous biologists failed to comprehend the significance of the laws of heredity for years after, and even today their significance has not penetrated to all branches of learning. Dr. Davenport understood at once the importance of Mendelism and espoused it with all the tremendous vigor he possessed. His main failing was his overenthusiasm for his cause—the importance of the gene to mankind. The vilification which he received is the usual reward for crusaders. (E.W. Reed & S. Reed, 1965, p. viii)

The Reeds go on to explain that Miss Sadie Deavitt and Miss Marie Curial were trained as field workers at the Eugenics Record Office and were then assigned to Faribault where, from 1911 to 1918, they collected family histories, did interviews, and constructed pedigrees on the original 289 subjects. The study was reopened in 1949, and the Reeds traced the descendants until 1965.

The book is an amazing document. It contains page after page of family charts that are simply elaborations and extensions of those used by Goddard and the early workers at the Eugenics Record Office. The manner in which people are described is reminiscent of the Kallikak study. And the grand leap of faith is there also—IQ scores and subjective judgments go unquestioned; poor performance and problems in adjustment are attributed to heredity.

In their introduction, the Reeds pay their respects to Goddard as the person who produced the "first family study of mental retardation we wish to mention"; they remark that the study was an "important and valuable contribution at the time" (E.W. Reed & S. Reed, 1965, p. 2). Then, after they discuss how their study was conducted, display their charts, and summarize their findings, they present some rather sweeping and startling conclusions:

We end our discussion with the perhaps euphoric opinion that the intelligence of the population is increasing slowly, and that greater protection of the retarded from reproduction will augment the rate of gain. The elevation of the average intelligence is essential for the comprehension of our increasingly complicated world. (p. 79)

The Reeds' view, then, was that institutionalization was resulting in an overall gain of intelligence in the population and that, as more defective people were institutionalized and prevented from reproducing, that gain would be increased. (It should be noted here that the Reeds reached these conclusions in a period when many mildly retarded people with no organic brain damage were still being kept in institutions and before deinstitutionalization had become a substantial force.)

In discussing the importance of their study, the Reeds felt that a significant "humanitarian" aspect of it was that it demonstrated that a better legal basis should be provided for the sterilization of "higher grade retardates" in the community. They even went beyond the argument that sterilization should be used to prevent hereditary problems: "Few people have emphasized that where the transmission of a trait is frequently from parent to offspring, sterilization will be effective and it is irrelevant whether the basis for the trait is genetic or environmental" (p. 77). In effect, they were advocating that the problem of environment as well as that of heredity could be resolved through sterilization: Both nature and nurture could be improved with this technique. Taken literally—and we have no reason to do otherwise— they were saying that the problem of poverty could be resolved by sterilizing the poor, the problem of ignorance could be remedied by sterilizing the ignorant—the applications seem limitless!

The Reeds' vision of the positive effects of sterilization was expressed in one of their summary statements: "When voluntary sterilization for the retarded becomes a part of the culture of the United States, we should expect a decrease of about 50 percent per generation in the number of retarded persons" (pp. 77–78). A staggering percentage decrease, if it were possible; and a seemingly easy, fast, and inexpensive way of "curing" a major social ill. Once again, however, it must be recognized that the Reeds' claim is an echo from 1912.

The extent to which the Kallikak myth survived the terrible truths of World War II and manifested itself in diverse areas of scientific and social thought is, at times, difficult to believe. One of the earliest books on ecology, *Road to Survival*, was written by William Vogt and published in 1948. Vogt had been much influenced by eugenic ideas; in his book he applied the ideas of the economist Thomas Malthus to natural resources and conservation. The central themes were that overpopulation posed the greatest threat to our exhaustable supply of natural resources and to economic and social stability worldwide. Vogt argued that, only by curbing population growth in "backward cultures" and in the lower classes of all societies, could humankind survive. He proposed harsh measures to control population. For example, he opposed foreign aid that would provide food and medical care to China or India. He felt that a high death rate in such countries was a "national asset." In short, he thought death should be allowed to do its work in bringing population growth under control (Vogt, 1948).

In a section of the book headed "Kallikaks of the Land," Vogt transposed eugenic thought to two occupational groups that he believed were ecologically and economically destructive—sheepherders and cattlemen:

The question of how to solve our forest problems opens up the wide, grim vista of ecological incompetence. The Jukeses and the Kallikaks—at least those who are obtrusively incompetent—we support as public charges. We do the same with the senile, the incurables, the insane, the paupers, and those who might be called ecological incompetents, such as the subsidized

stockmen and sheepherders. These last, in so far as they
deteriorate and destroy the grasses, expedite erosion,
and contribute to flood peaks, are worse than paupers.
They exist by destroying the means of national survival;
were we really intelligent about our future, we would
recognize such people as Typhoid Marys—the source of
environmental sickness with which they are infecting us
all. . . . In our national interest they must be liquidated,
at least in part. In the process, a good many people are
certain to be hurt, as in any liquidation. But the longer it
is postponed the more people will suffer. (Vogt, 1948,
p. 145)

Assuming that Vogt was speaking only of the "liquidation"
of the jobs of these people, not the people themselves, it is an
interesting twist of the old eugenic idea that at times social
action must be taken toward certain types or classes of
human beings. Vogt also, however, held to some of the more
traditional concepts. On the issue of sterilization, for example, he observed:

There is more than little merit in the suggestion . . . of
small but adequate amounts of money to be paid to
anyone—especially the males—who would agree to
the simple sterilization operation. . . . Since such a
bonus would appeal primarily to the world's shiftless, it
would probably have a favorable selective influence.
From the point of view of society, it would certainly be
preferable to pay permanently indigent individuals,
many of whom would be physically and psychologi-
cally marginal, $50 or $100 rather than support their
hordes of offspring that, by both genetic and social
inheritance, would tend to perpetuate the fickleness.
(Vogt, 1948, p. 145)

In 1956, William Shockley was awarded the Nobel Prize
for Physics. The award came as a result of his work with Bell
Laboratories in the development of the transistor. Shockley

has held teaching and research positions in several major universities and is presently on the faculty at Stanford. Today, however, he is known less for his accomplishments in physics than for his views on race and intelligence.

Shockley argues that intelligence is largely hereditary and that the black race is innately inferior in intellect. Like many of the early eugenicists, Shockley has no credentials in genetics, nor does he have a background in the social sciences. He simply began to voice his views on intelligence, genetics, and race; and many people have listened receptively.

In 1966, in a lecture at Stanford University, Shockley stated that the Kallikak study should not be dismissed lightly, that the

bad hereditary concept may have been too enthusiastically rejected by perfectionists. . . . Can it be that our humanitarian welfare programs have already selectively emphasized high and irresponsible rates of reproduction to produce a socially relatively unadaptable human strain? (Chase, 1977, p. 158)

In the mid seventies Shockley proposed a "thinking exercise" about sterilization. He hoped the exercise would stimulate thinking about dealing with the problems created by inherited defects in intelligence. His plan involved the award of cash bonuses to people who scored low on intelligence tests and agreed to be sterilized:

At a bonus rate of $1,000 for each point below 100 I.Q., $30,000 put in trust for a 70 I.Q. moron potentially capable of producing 20 children might return $250,000 to taxpayers in reduced costs of mental retardation care. Ten percent of the bonus in spot cash might put our national talent for entrepreneurship into action. (Shockley, 1976, p. 166)

Note that Shockley's connection with old-line eugenics extends even to using the term moron. Most other contemporary hereditarians seem to avoid identification, at least publicly, with what they might consider the historical excesses of the movement. Shockley speaks unabashedly of eugenics, dysgenics, and morons.

In his sterilization proposal, Shockley anticipates the problem of reaching those who most need to have the surgery done:

> A feature that might frustrate the plan is that those who are not bright enough to learn of the bonus on their own are the ones most important to reach. The problem of reaching such people is what might be solved by paying the 10 percent of the bonus in spot cash.
>
> Bounty hunters attracted by getting a cut of the bonus might then persuade low I.Q., high-bonus types to volunteer. (Shockley, 1976, p. 166)

In a 1980 interview, Shockley discussed his donation to the Nobel-laureate sperm bank established by eugenics advocate Robert Graham. In defending his participation and the reasonableness of Graham's project, Shockley explained:

> Graham's interest in the declining quality of people goes back at least to the Sixties, when he wrote a book called *The Future of Man*. He did studies of what went on during the French Revolution and the elimination of the elite class, which probably removed some of the brilliant people of France. I don't know that one can say France has significantly less intellectual potential now than it did before the Revolution, but this is what some of Graham's studies were concerned with. Anyway, Graham had for some time been urging more intelligent people to have more children. We had talked about these things and my concern about possible downbreeding, or dysgenics, struck a responsive chord in

him. I knew about his plans for a sperm bank and when it was set up, I had no particular problem in making a decision. (Jones, 1980, p. 72)

The Nobel-laureate sperm bank received a great deal of attention from the press at the time of its establishment. Much less has been heard of it recently. The idea of creating "superhumans" by inseminating outstanding female human specimens with the sperm of men who have demonstrated their superiority by receiving Nobel prizes is an esoteric idea but clearly in keeping with the eugenic tradition. Shockley was obviously a dedicated contributor to the bank and to the tradition.

Graham's book, *The Future of Man,* which Shockley cited, contains a passage that seems to embody the rationale for establishing the sperm bank:

> We may lift our eyes unto lofty goals ahead but the path toward them is to be trod one step at a time. Nor is ascent as easy as descent. Is it possible, then, voluntarily and within our laws and mores, to put intelligent selection to work for the sake of man? There are indeed many good ways to do so.
>
> The plenary solution to the great problem described in this book, and the essence of intelligent selection is for the intelligent to release much of their natural fertility which they have repressed so long, and at the same time assist the mentally deficient voluntarily to reduce their output of offspring. (Graham, 1970, p. 157)

In the earlier-cited interview with Shockley, he refers to Elmer Pendell, the demographer, who argues that civilizations decline when "problem makers" multiply at a greater rate than "problem solvers." In his book, *Sex Versus Civilization,* Pendell proposes a law that would prevent the marriage and reproduction of problem makers. Among those not allowed to marry unless sterilized would be "those who cannot earn a living" and "those of very low IQs or less than four years of education" (Pendell, 1967, p. 198). The prohibition imposed by the law, even its wording, are strikingly similar to the Virginia miscegenation law of the 1920s and the German law of the 1930s forbidding marriage between Jews and Aryans. The philosophical intentions are the same: the preservation of the "quality" of the culture and the prevention of degeneracy.

In his book, Pendell illustrates the potential power of an idea combined with a teacher. In a college class he taught on population, he covered various propositions for controlling the growth of a population, or certain parts of a population. These included sterilization, limitations on immigration, and restrictive marriage laws. In the spring of 1965, near the end of the semester, Pendell handed out ballots and gave each student the opportunity to vote for or against each of these measures. The results he reported are sobering; 88 percent of the students were in favor of increased restrictions on immigration, 76 percent were in favor of laws restricting marriage, and 73 percent supported the idea of a bonus plan to encourage sterilization (Pendell, 1967, pp. 204–205). This is another example of the power, appeal, and persistence of eugenic concepts and the seeming social amnesia regarding the tragedies they have engendered in the past.

Eugenic programs have often been proposed in ways that make them appear to be not only for the good of society but also for the good of the victim. I suspect that in some cases the proposers of such measures sincerely believed that they were looking out for the best interests of the inferior. In any case, from the Kallikaks onward, eugenic actions have often been presented as being "for their own good": The retarded should be institutionalized for their own protection. They should be sterilized so that they can be released. They should be allowed a "good death" so that they do not have to bear the pain of a "life devoid of meaning."

In this vein, here are William Shockley's views on the inferiority of black people:

The phrase that I now use is the The Tragedy for American Negroes. My emphasis is on the tragedy for the Negroes themselves arising from their greater per-capita representation in statistics for poverty, welfare, educational failure and crimes. The relief burden related to these statistics could be called a National Negro Tragedy if the intent is to focus upon the concerns of taxpaying citizens. But that is an unfair focus. I believe society has a moral obligation to diagnose the tragedy for American Negroes of their statistical I.Q. deficit. Furthermore, this is a worldwide tragedy, and in my opinion, the evidence is unmistakable that there is a basic, across-the-board genetic disadvantage in terms of capacity to develop intelligence and build societies on the part of the Negro races throughout the world. (Jones, 1980, p. 81)

Shockley states repeatedly that his aim is to limit human misery. He never, however, addresses the fact that the means by which he proposes to reach that aim have produced human misery on an unbelievable scale. He insists that he believes in the equality of people:

Yes, I believe in the created equal assertion of the Declaration of Independence, when it is interpreted in terms of equal political rights, but I would qualify it some: I don't think the right should be given equally to everyone to have children, if those people having children are clearly destined to produce retarded or defective children. (Jones, 1980, p. 98)

William Shockley has enjoyed considerable popularity, and his ideas have been accepted in certain sectors of the general population and, of course, among those special interest groups who find his arguments appealing and useful to their causes. He has not, however, had much impact in academic circles. In fact, he has been criticized and his ideas have been repudiated by members of the faculty teaching genetics at his own university. Indeed, he has been asked *not* to speak on eugenics by several professional organizations who recognize his contributions as a physicist but not as a geneticist or psychologist.

The work of Professor Arthur Jensen is different in tone and style from that of Shockley. Rather than advocating specific eugenic solutions, such as sterilization, to the problem of low intelligence, Jensen has concentrated his efforts on building a statistical case for the heritability of mental ability. He has done so in a careful, scholarly fashion and has, therefore, enjoyed acceptance in some quarters of the academic community. Because of this greater receptivity at higher levels of influence, Jensen's work has probably had more impact on public attitudes and policies than has Shockley's.

In 1969, the *Harvard Educational Review* published Jensen's now famous article, "How Much Can We Boost IQ and Scholastic Achievement?" In the article, Jensen maintained that compensatory education programs aimed at minority and poverty populations had failed to make any difference. The reason, he argued, is that I.Q. is hereditary and fixed—it cannot be significantly changed through educational intervention. The major points of his article have been summarized by Chorover:

1. Scholastic achievement (success in school) depends upon mental ability (commonly called "intelligence").

2. Intelligence is a complex trait and is difficult to define but it can be measured independently by performance on IQ tests.

3. IQ test scores correlate strongly with scholastic success, family income, parents' occupational status, and other sociocultural indices.

4. Differences in IQ scores (whether between individuals or groups) are mainly attributable to genetic factors.

5. Because the differences in mental ability responsible for the differences in IQ test performance and scholastic achievement are attributable to genetic factors, efforts to "boost" them have been (and must forever be) largely unsuccessful. (Chorover, 1979, p. 31)

Jensen's claims here and in subsequent publications rest largely upon his detailed reviews of studies done over the years using intelligence tests with minority groups, particularly with black subjects. His analyses of these studies have led him to assert that the differences found between blacks and whites on intelligence tests is hereditary in origin. Further, he believes that black people on the average inherit a lesser capacity for dealing with abstractions and symbolic material.

Jensen's work is meticulous and sophisticated. To many who have examined his analyses, their scientific precision is either convincing or intimidating, or both. It is thus helpful to look at what Jensen presents with an eye to separating what he says from how he says it. What he is saying is not new; it amply reflects the influence of eugenics on his thinking. Examination of two examples of such influence may serve to clarify the nature of Jensen's work.

In his book, *Genetics and Education,* Jensen gives credit to Audrey Shuey for having produced the most comprehensive review of studies of the intelligence of black subjects. Having analyzed and reported on 382 such studies, Jensen observes that

the basic data are well known: on the average, Negroes test about 1 standard deviation (15 IQ points) below the average of the white population in IQ, and this finding is fairly uniform across the 81 different tests of intellectual ability used in the studies reviewed by Shuey. (Jensen, 1972, p. 161)

Jensen cites extensively from Shuey's work. In fact, he bases his estimate of the IQ gap between black and white Americans on Shuey's earlier findings—an average IQ of 100 for whites and an average IQ of 85 for blacks. In discussing the causes of this gap, Jensen says:

> In view of all the most relevant evidence which I have examined, the most tenable hypothesis, in my judgment is that genetic, as well as environmental, differences are involved in the average disparity between American Negroes and whites in intelligence and educability, as here defined. All the major facts would seem to be comprehended quite well by the hypothesis that something between one-half and three-fourths of the average IQ difference . . . is attributable to genetic factors, and the remainder to environmental factors and their interaction with genetic differences. (Jensen, 1973, p. 363)

Audrey Shuey was a professor and chairman of the Department of Psychology at Randolph-Macon Women's College in Lynchburg, Virginia. Her book, the one referred to so often by Jensen, is entitled *The Testing of Negro Intelligence*. It was first printed in 1958; a second edition was published in 1966. Shuey's concluding statement in the second edition clearly indicates the substance and significance of her study:

> The remarkable consistency in test results, whether they pertain to school or preschool children, to children between Ages 6 to 9 or 10 to 12, to children in Grades 1 to 3 or 4 to 7, to high school or college students, to enlisted men or officers in training in the Armed Forces—in

World War I, World War II, or the Post-Korean period—
to veterans of the Armed Forces, to homeless men or
transients, to gifted or mentally deficient, to delinquent
or criminal; the fact that differences between colored
and white are present not only in the rural and urban
South, but in the Border and Northern states; the fact
that the colored preschool, school, and high school
pupils living in Northern cities tested as far below the
Southern urban white children as they did below the
whites in the Northern cities; the fact that relatively
small average differences were found between the IQ's
of Northern-born and Southern-born Negro children in
Northern cities; the fact that Negro school children and
high school pupils have achieved average IQ's slightly
lower in the past twenty years than between 1921 and
1944; . . . all taken together, inevitably point to the presence of native differences between Negroes and whites
as determined by intelligence tests. (Shuey, 1961,
pp. 520–521)

Audrey Shuey did her graduate study in psychology at
Columbia University. She received her doctorate there in
1929. During her years at Columbia, she studied and
worked with Henry Garrett. The foreword to her book was
written by Garrett. In it, Garrett states:

The question of Negro-white differences in mental test
performance has been the subject of lively debate in
recent years. Unfortunately, the subject has often been
confused with social and political issues of racial
inferiority, desegregation, civil rights and other extra-
neous matters. Moreover, a number of well meaning
but often insufficiently informed writers have taken the
untenable position that racial differences ought not to
be found; or if found should immediately be explained
away as somehow immoral and reprehensible. With
this attitude I am in sharp disagreement. I welcome

every honest effort to help Negroes improve their lot, but I do not believe it is necessary to "prove" that no racial differences exist, nor to conceal and gloss them over, if found, in order to justify a fair policy toward Negroes. The honest psychologist, like any true scientist, should have no preconceived racial bias. He should not care which race, if any, is superior in intelligence, nor should he demand that all races be potentially equal. He is interested simply in uncovering differences in performance when such exist and in inferring the origin of these differences. And this is certainly a legitimate scientific enterprise. (Shuey, 1966, p. vii)

It is ironic that the strongest advocate of the Kallikak myth at that time and the promoter of the "breeding-down" miscegenation warning should speak of no "preconceived racial bias" as the hallmark of an honest psychologist.

Garrett, who by this time had retired from Columbia and was a visiting professor at the University of Virginia, concluded his remarks in the foreword to Shuey's book with these words: "We are forced to conclude that the regularity and consistency of these results strongly suggest a genetic basis for the differences. I believe that the weight of evidence (biological, historical and social) supports this judgment" (p. viii).

Another person who was influential in the development of Arthur Jensen's thinking and techniques was Sir Cyril Burt. A British psychologist, Burt was an enthusiastic investigator of the concept of the genetic transmission of mental traits. He believed that low IQ and poor academic performance were the result of "inborn inferiority of general intelligence" (Chorover, 1979, p. 49). Burt was best known for his studies of identical twins who were raised apart. His data showed that, even when twins were raised in very disparate environments, their IQs were not significantly different. According to Burt, this was evidence that the genetic inheritance of the twins, not the environment in which they were raised, determined their intelligence.

In *Genetics and Education,* Jensen cites Burt's work more than that of any other source except himself. In fact, Jensen explains, it was Burt who first inspired his interest in the inheritance of intelligence:

> While in London, I had had the privilege of attending the Walter Van Dyke Bingham Memorial Lecture, sponsored by the American Psychological Association, and delivered that year (May 21, 1967) by Professor Sir Cyril Burt, whose topic was "The Inheritance of Mental Ability." I did not go to the lecture out of any special interest in the topic but simply because Sir Cyril Burt, who was then in his seventies, was one of England's most famous psychologists, and I merely wanted to see him in person. His lecture was impressive indeed; it was probably the best lecture I ever heard, and I recommend it to all students of psychology and education. . . .
>
> So in preparation for writing the one chapter of my book on the culturally disadvantaged that was to deal forthrightly with the genetics of intelligence, rather than ignore the subject or dismiss it cavalierly as so many writings in this field had done, I began by reading Burt's masterful Bingham Lecture, which led me to all his other excellent articles in this area, and soon I found myself engrossed in reviewing the total world literature on the genetics of human abilities. (Jensen, 1972, pp. 8–9)

Later in the book, Jensen praises Burt as the most distinguished exponent of the study of the heritability of intelligence and says that his writings are a "must" for all students of individual differences. Jensen saves his greatest praise, however, for Burt's studies of twins, and he reports on them extensively. That Burt was a central figure in Jensen's thinking is apparent throughout the latter's works.

After Burt's death, Professor Leon Kamin of Princeton reviewed the data from the studies of twins (Kamin, 1976):

Correlations Reported by Cyril Burt in His Studies of Twins

Year of Report	Identical Twins Raised Apart		Identical Twins Raised Together	
	No. of Pairs	IQ Correlation	No. of Pairs	IQ Correlation
1955	21	.771	83	.944
1958	"over 30"	.771	Not Available	.944
1966	53	.771	95	.944

Adapted with permission from *The Science and Politics of I.Q.* by L. Kamin (p. 38), 1974, Potomac, Md.: Lawrence Erlbaum Associates.

Kamin noticed that, although the number of pairs of twins that Burt used in the studies varied, the correlations he reported between the intelligence test scores of twins raised apart remained the same. The chances of this kind of statistical consistency occurring in separate studies using different sample sizes is infinitesimal. Kamin looked more closely at Burt's work and discovered that this sort of statistical constancy was to be found throughout the research. Kamin finally came to the conclusion that Burt's work on the twins simply could not be accepted with any confidence in its scientific validity. In 1976, Oliver Gillie, a medical reporter for the London *Sunday Times* stated that he had found evidence that the assistants Burt said had worked with him on the twin studies never existed. Gillie said that the people who Burt claimed had seen the raw data from the studies, had helped him perform the statistical analyses, and who were listed as coauthors of the final reports either never existed or could not have been in contact with Burt when the work was done. In his biography of Cyril Burt, L.S. Hearnshaw corroborated the charge that Burt had doctored the findings in the twin studies and had created mythical assistants. He also found other instances of fraud and distortion in Burt's work (Hearnshaw, 1979). Thus, one of the cornerstones of Arthur Jensen's work was discredited.

In fact, Jensen's work is based squarely on the same concepts that were central to Goddard's research and his interpretation of his findings: Intelligence tests yield a valid measure of intellectual capacity, and differences in intelligence as measured by the tests are innate. Seymour Sarason has referred to Jensen's attachment to standardized testing as the "Achilles heel" of his position (Sarason, 1984, p. 21). Regardless of Jensen's meticulous methods and statistical accuracy, if his assumptions about the validity of test results and the unmodifiable nature of intelligence are weak, the whole edifice based on his data crumbles, or should.

Leon Kamin has called the existing literature on the heritability of IQ disgraceful. He maintains that there is no convincing evidence of the genetic nature of intelligence. He goes on to say:

> The conceptual and empirical errors of the "scientists" of IQ and heritability have done real mischief. They may be capital fellows personally, but the objective consequence of their invalid work is a furtherance of racism and of class exploitation. (Kamin, 1975, p. 491)

Jensen has never claimed that he has the final word on the nature-nurture question. He does, however, argue for the strength of his hypotheses concerning intelligence and for the validity of the data he uses to support those hypotheses. His assertions that "scientific evidence" indicates that there are racial differences in the genetic endowment of intelligence thus have had a profound impact.

From Goddard to Jensen, the designation "scientific" has lent a credibility to arguments that have influenced the thinking and behavior of people far removed from the actual research studies with all of their limitations and weaknesses. For generations, teachers who must decide what is possible to accomplish with their students, politicians who make decisions about the most effective allocation of funds, and racists who seek validation of their prejudices have all been influenced and bolstered in their opinions by the eugenic conceptual tradition. The name, the religion, and the race may vary, but the Kallikaks are still being hunted, found, and blamed for the ills of society.

REFERENCES

Chase, A. (1977). *The legacy of Malthus: The social costs of the new scientific racism.* New York: Knopf.

Chorover, S. (1979). *From Genesis to genocide.* Cambridge, Mass.: MIT Press.

Garrett, H. (1955). *General psychology* (rev. ed.). New York: American Book Co.

Gillie, O. (1976, October 24). Criminal data was faked by eminent psychologist. *Sunday Times,* London, pp. 1–2.

Graham, R. (1970). *The future of man.* North Quincy, Mass.: Christopher.

Hearnshaw, L.S. (1979). *Cyril Burt, psychologist.* London: Hodder and Stoughton.

Jensen, A.R. (1972). *Genetics and education.* New York: Harper & Row.

Jensen, A.R. (1973). *Educability and group differences.* New York: Harper & Row.

Jones, S. (1980). Interview with William Shockley. *Playboy.* pp. 69–102.

Kamin, L. (1974). *The Science and politics of I.Q.* Potomac, Md.: Lawrence Erlbaum Associates.

Kamin, L. (1975). Reply to Samelson. *Social Research, 42,* 480–492.

Pendell, E. (1967). *Sex versus civilization.* Los Angeles, Calif.: Noontide Press.

Reed, E.W., & Reed, S. (1965). *Mental retardation: A family study.* Philadelphia: W.G. Saunders.

Sarason, S. (1984). Unlearning and learning. In B. Blatt & R. Morris, (Eds.), *Perspectives in special education: Personal orientations.* Glenview, Ill.: Scott, Foresman & Co.

Shockley, W. (1976). Sterilization—A thinking exercise. In C. Bajema, (Ed.), *Eugenics: Then and now.* New York: Halsted Press.

Shuey, A. (1966). *The testing of Negro intelligence* (2nd ed.). New York: Social Science Press.

Vogt, W. (1948). *Road to survival.* New York: William Sloane Associates.

Epilogue

I N THE MANUSCRIPT DIVISION OF THE LIBRARY OF CONGRESS
I found a letter from Henry Goddard to Arnold Gesell.
Gesell, of course, is remembered for his work on child development. Goddard apparently had known Gesell since they
were students together at Clark University. Gesell visited
with Goddard at Vineland in the summer of 1909, and he
later gave credit to that visit for stimulating his interest in
child psychology. Goddard's letter contains a reference to
an honorarium. This is in relation to an earlier letter from
Gesell in which he declines an informal offer to make a
presentation because he felt the honorarium offered was not
sufficient. The Goddard letter is dated December 30, 1928:

191

My Dear Gesell:

Your good letter of last November was duly received and you have long since guessed the answer. I fully appreciate and agree with your position. Now I can save you the annoyance of having to decline an honorarium that really does not meet the situation. Whenever we can reach a figure that makes it worthwhile I will let them call you. Until then I [will] simply say you find it hard to leave your work, and it is no use asking you.

I am sorry that we could not get to New York and see you and Mrs. Gesell, but it was impossible. Mrs. Goddard had a mild attack of the flu and I had to be nurse and cook. She is better now, so that I am going to run up to Battle Creek next Tuesday to "preach" a little at the Eugenics Conference. You see Kellogg entertains the whole association and there is just enough Scotch or Jew in me that I could not miss a free dinner! . . .

We must somehow get together and have a powwow and gabfest before very long. I am off duty the spring quarter also. Have not yet decided what to do. We should start west at once but again that Jewish blood (I think it is Jewish—look at the nose!) makes me wait until the summer excursion tickets are on sale—May 15th.

Well, if I cannot tell you what we are going to do I will tell you later what we did do.

Yours with the Goddards' best wishes for a Happy New Year for all the Gesells.

"Herbert"

I am including this letter here after previously deciding not to make reference to it. I thought earlier that it was not important enough to mention, that it might be construed as an unnecessary slap at Goddard's character, and that it would serve no purpose. I have come, however, to conclude just the opposite—that it serves a very important and instructive purpose. Goddard, in writing to a professional friend and colleague, shows no hesitation in making remarks that can only be described as slurs against Jews. He obviously has no reluctance in making these defamatory references in correspondence with one of the most influential and respected social scientists of the time. Perhaps Goddard's words would not have stood out as unusual or inappropriate in academic and professional circles in 1928. Unfortunately, they would probably still be acceptable in some of those same circles today. Goddard, I am certain, would have seen no great harm in what he jokingly said about Jews. He would have had great difficulty in seeing how the stereotypes expressed in his letter would be one of the seeds that led to the Holocaust. He would also never see the connection between the Kallikak myth he created and the needless sterilization of thousands of Americans or the German race hygiene program.

I do not believe that Henry Goddard was a sinister man intent upon doing harm to the poor, the foreign-born, the uneducated, or people with different racial or religious backgrounds. He was as much a product of a powerful idea as he was the creator of a social myth. He took the idea, cast it with characters, and embellished it with stories of what he wanted to be true of the characters—he saw what he wanted to see in the Kallikaks. The idea that shaped Goddard and from which he shaped his story has produced tragic results: It has perpetuated the idea that some families, some nationalities, some races, some religious groups, some social classes are naturally, inherently, and unmodifiably inferior.

In this book, I have attempted to describe the making of a social myth and to illustrate how lives were restricted, damaged, and even destroyed as a result of that myth. In the process of researching and writing it, I have been reminded of, and made more sensitive to, how careful we must be in the sciences and in human service professions about the myths that we accept, foster, or even create. Myths have a way of becoming reality. Myths have a way of gathering force as they are passed along. They have a way of surviving the intent and lifetime of their creators.

REFERENCES

Goddard, H.H. (1982). [Letter to Arnold Gesell]. *Gesell papers.* Washington, D.C.: Library of Congress, Manuscript Division.

Index

A

Abnormal Psychology, 67
Act to Preserve Racial Integrity
  of 1924, Virginia, 156, 157
Allen, Garland, 167
Allied Jewish Appeal, 45
American Association for the
  Advancement of Science, 69
American Catholic Historical
  Association, 59
American faith in education, 1
American Institute for Mental
  Studies, 86
  See also Training School
  for Feeble-Minded Girls
  and Boys, Vineland, New
  Jersey
American intellect, 126

*American Journal of
  Psychology*, 62
American Psychological
  Association, 170, 187
Anti-Jewish attitudes of
  Goddard, 192, 193
Antioch College Research
  Institute of Human
  Development, 45
Applied Eugenics, 136
Archives of the History of
  American Psychology,
  University of Akron, 51, 73,
  132, 160
Aristocracy, 128, 129, 130
Army Alpha, 126
Army Beta, 126
Army tests, 126, 127, 132
Association for Voluntary
  Sterilization, 158

Average intellectual level in
    American population, 126

B

"Bad seed" concept, 21
Beard, Virginia, 144
Bell, Alexander Graham, 42
Bell Laboratories, 176
Besondere Heilverfahren
    ("special healing
    procedures"), 164
Binding, K., 165
Binet, Alfred, 41, 56
Binet-Simon Test, 56, 139–140
Binet tests, 41, 117, 118
    of immigrants, 119
Bird, Randall, 167
Blacks
    inferiority of, 180
    as inmates, 135
    intelligence of, 177, 184, 185
    and intelligence tests, 183
Booker, B., 156
Breeding Down, 170
"Breeding down" (dysgenics),
    172, 178, 186
Brigham, C.C., 126, 127
Broadway play proposals for
    Kallikak book, 62, 63, 64
Buck v. Bell, 139, 149, 150, 151,
    156
Buck, Carrie, 139, 140, 142, 143,
    144, 145, 146, 147, 148, 149,
    150, 152
Buck, Doris, 144, 150
Buck v. Priddy, 141, 142, 143,
    144, 145, 147, 148, 149
Buck, Vivian, 152
Bulletin No. 1, Eugenics Record
    Office, 143

Bulletin No. 2, Eugenics Record
    Office, 42–43
Burbank, Luther, 42
Burt, Cyril, 186, 187, 188
Busing controversy, 172

C

Canot, Robert, 162
Carnegie Institute, 41–42, 43,
    44, 157
Central Virginia Training
    Center, 150
Chase, Allan, 139, 161, 170, 177
Chicago Daily News, 161
Child labor restriction, 136
Chorover, Stephen, 170, 172,
    182, 183, 186
Clark University, 38, 191
Collier's Weekly, 131
Colonization, 19
Columbia University, 185, 186
    Department of Psychology
        of, 170
    Teachers College of, 132
Committee on Provision for the
    Feeble-Minded, 42
Compulsory education, 136
Compulsory sterilization, 3, 7,
    19, 38, 63, 137–165
    and informants, 138–139
    in Nazi Germany, 162
    U.S. Supreme Court ruling
        on, 139, 149, 154
Conklin, 67
Conot, R.E., 164
Conservatism of Goddard, 132
Control of human
    reproduction, 42, 43, 131,
    136, 137
Cremin, L., 2

Criminals' intelligence, 135
Crissey, Marie Skodak, 38, 41, 42, 43
Curial, Marie, 173

D

Darwin, Charles, 2
Davenport, Charles B., 42, 43, 81, 138, 172, 173
Deats, Hiram, 98, 99
Deavitt, Sadie, 173
Dedication of Kallikak book, 45
DeJarnette, J.S., 140, 141, 142
Democracy, 7
   Goddard's views on, 128, 129, 130
Deportation of immigrants, 7
Desegregation of races, 170, 172
Detamore, Carrie Buck, 151
   See also Buck, Carrie
Detection of morons, 117
Development of Intelligence in Children, The, 56
Devery, E.D., 38, 57
Dial, The, 62
Die Famile Kallikak, 161, 167
Dobbs, Alice, 151, 152
Dobbs, J.T., 151, 152
Dobbs, Vivian, 152
Dodd Mead & Co., 63
Doll, Edgar, 27, 73, 74, 75, 131, 160
Doll, Eugene, 27, 29, 30, 31, 35
"Down breeding" (dysgenics), 172, 178, 186
Dudley, Richard, 145
Dudley, Samuel, 146
Duell, Wallace, 161
Dunlap, Knight, 68, 69, 70

Dysgenics ("down breeding"), 172, 178, 186

E

East, E.M., 65, 66
Eastern Europeans, 127, 138
Education
   for all Americans, 126
   American faith in, 1
   compulsory, 136
   Goddard's views on, 7, 129, 130, 131
Ellis Island, 6, 116, 117, 118, 119
   See also Immigrants;
   Immigration
Emotional components of Deborah's problems, 26
Environment, 124
Equality of people, 181
Estabrook, Arthur H., 142, 143, 145, 146, 147, 156, 157
Ethnic inferiority theories, 2
Eugenical News, 156
Eugenics, 2, 3, 4, 7, 8, 9, 42, 43, 61, 62, 128, 137, 156, 157, 158, 159, 160, 162, 165, 169, 175, 176, 178, 179, 180, 182, 183, 189–190
   Nazi, 165
   new, 8, 169–190
Eugenics Conference, 192
Eugenics Record Office of Cold Spring Harbor, 3, 42, 43, 137, 138, 139, 148, 156, 172, 173
   Bulletin No. 1 of, 143
   Bulletin No. 2 of, 42–43
Eugenics in Relation to the New Family, 156–157
Euthanasia programs in Germany, 164, 165

F

Faribault State School and
Colony, Minnesota, 172, 173
Federation of Jewish Charities,
45
Feeble-Mindedness: Its Causes
and Consequences, 13, 65
Fels and Company, 45
Fels, Samuel S., 44, 45, 50
Fernald, Dr., 71
Fiedler, James T., 57
Figgins, Doris Buck, 150, 151
See also Buck, Doris
Figgins, Matthew, 150
"Final solution" of Hitler, 8
"First method" (F.M.), 120
F.M. See "First method"
Fogel, Martin, 162
Friedrich Mann's
Pedagogisches Magazin, 161
From Genesis to Genocide, 170
Future of Man, The, 178, 179

G

Galton, Francis, 2
Garrett, Henry, 170, 172, 185,
186
Garrison, S. Olin, 37, 38
Genealogical Magazine of
New Jersey, The, 98
General Psychology, 170
Genetics, 12
defects in as threat to society,
42
and feeblemindedness, 43,
66
and intelligence, 126, 187,
189
and mental ability, 182
and mental defects, 21
and mental retardation, 3, 4,
11, 74
unit trait, 65–66, 75
Genetics and Education, 183,
187
Genocide by Nazis, 166
German Criminal Code, 159
German edition of Kallikak
book, 161, 167
German Hereditary Health
Law, 154, 155, 156
German Hygienic Museum,
162
German race hygiene
program, 7, 156, 159
German sterilization programs,
158, 160
Gesell, Arnold, 191, 192
Gillie, Oliver, 188
Goddard, Henry Herbert, 3, 4,
5, 6, 8, 11, 12, 13, 15, 16, 17,
18, 19, 21, 22, 23, 26, 27, 30,
33, 34, 35, 38, 39, 40, 41, 43,
44, 45, 49, 50, 51, 55, 56, 61,
62, 63, 64, 65, 66, 67, 69, 70,
71, 72, 73, 74, 75, 76, 77, 78,
79, 83, 84, 85, 86, 87, 88, 89,
90, 91, 93, 97, 98, 100, 101,
102, 103, 104, 106, 107, 109,
110, 116, 117, 118, 119, 120,
121, 122, 123, 124, 125, 126,
127, 128, 129, 131, 132, 133,
134, 135, 136, 137, 142, 143,
159, 160, 161, 162, 172, 173,
188, 189, 191
anti-Jewish attitudes of, 192,
193
conservatism of, 132
and democracy, 7, 128, 129,
130
and education, 7, 129, 130,
131

and immigration, 6
research strategy of, 12
and voting, 130
Gould, Stephen Jay, 79, 80, 83,
128, 151, 152, 153
Graham, Robert, 178, 179

H

Hackett, F., 165, 166
Haller, M.H., 42, 46
Hall, G. Stanley, 38, 39, 41
Handicapped person killings in
Germany, 166
Harlow, Richard, 146
Harriman, E.H., Mrs., 42
Harriman, W. Averell, 42
Harvard Educational Review,
182
Haverford College, 39
Hearnshaw, L.S., 188
Hebrew Immigrant Aid
Society, 45
Heidelberg University, 157, 158
Hereditary Health Law,
Germany, 154, 155, 156
Heredity. *See* Genetics
Herrnstein, R., 127
"Higher grade retardates," 174
"High grade defectives," 12
Hitler, Adolf, 154, 156, 164, 165
"final solution" of, 8
Hoche, A., 165
Hoehn, M., 50, 59
Holmes, Oliver Wendell, 149,
151
Holmes, S.J., 156
Holocaust, 7, 128, 166, 169, 193
Hopkins, Charles, 64, 65
Hopkins, John W., 145, 146

House Committee on
Immigration and
Naturalization, 138
"How Much Can We Boost IQ
and Scholastic
Achievement?," 182
Human pedigrees, 3
Human reproduction control,
42, 43, 131, 136, 137
Hungarians, 118, 119, 127
Hunterdon County Historical
Society, 98

I

Identical-twin studies, 186, 187,
188
Idiots, 41
Imbeciles, 41
Immigrants, 116
See also Ellis Island
deportation of, 7
intelligence of, 6, 45
Jewish, 45, 118, 121
testing of, 45, 119
Immigration, 6, 115, 127, 138
public opinion about, 127
restricted, 3, 7, 180
from southern Europe, 128
Immigration Restriction Act of
1924, 3, 7, 45, 127
Incurability of morons, 34
Incurable-person mercy
killings in Germany, 164
Independent, The, 62
Indiana State School for the
Feeble-Minded, 38
Inferiority of blacks, 2, 180
Informants under sterilization
laws, 138–139
Inheritance. *See* Genetics

Inheritance of Mental Disease,
  The, 66
Institutionalization, 7, 41, 137,
  174
  of Deborah, 26, 27
  segregation by, 19
Integration, 170, 172
Intellectual aristocracy, 129
Intelligence, 8, 182
  See also IQ; Mental ability
  of American population, 126
  of criminals, 135
  and genetics, 126, 187
  of immigrants, 6
  and race, 177
Intelligence of the
  Feebleminded, The, 56
Intelligence tests, 33, 41, 56,
  117, 136, 183, 185
  See also specific tests
  of immigrants, 45
  and Nazis, 161
Intermarriage, 172
IQ, 182, 183
  See also Intelligence
  of blacks, 184, 185
  and genetics, 189
IQ and Racial Differences, 172
Italians, 118, 119, 120, 127

Jones, S., 179, 181
Journal of Heredity, 75
Journal of the History of
  Biology, 167
Journal of Psycho-Asthenics,
  130–131
Justice at Nuremberg, 162

K

Kaempeffert, Waldemer, 70
Kallikak Family: A Study in the
  Heredity of Feeble-
  Mindedness, The, 3, 13
  Broadway play proposals
    for, 62, 63, 64
  dedication of, 45
  German edition of, 161, 167
  photo retouching in, 79, 80,
    83–84
  popular appeal of, 62
  reviews of, 62, 65, 68, 75
Kamin, Leon, 188, 189
Kauser, Alice, 63, 64
Kellogg, John H., 81, 192
Kite, Elizabeth S., 4, 5, 13, 15,
  16, 17, 49–59, 67, 72, 91, 98,
  122, 123, 160
Kopp, Marie, 160

J

James, D.L., 64, 65
Jensen, Arthur, 182, 183, 184,
  186, 187, 188, 189
Jewish immigrants, 45, 118,
  120, 121, 127
  testing of, 119
Jewish refugees, 128
Johnson, R., 136, 137
Johnstone, Edward R., 38, 39,
  43, 64, 116, 129

L

Landis, Virginia, 144
Language-related difficulties,
  26
Laughlin Harry H., 137–138,
  139, 140, 142, 143, 151, 154,
  157, 158, 167
Law for the Protection of
  German Blood and Honor,
  155

Laws
  See also specific laws
  immigration. *See*
    Immigration Restriction Act
    of 1924
  sterilization. *See* Compulsory
    sterilization
Lazarus, Emma, 115, 116
Learning disabilities, 23, 26
Life, 164
Lippmann, Walter, 127
Lombardo, Paul A., 151, 152
London *Sunday Times*, 188
Ludmerer, K.M., 155, 156
Lynchburg College, Virginia,
  167

M

Malthus, Thomas, 175
Marks, R., 139
Marriage
  among feeble-minded, 38
  prevention of, 131
  between races, 172
  restrictive laws on, 180
  selective, 2
Maxham, Benjamin, 37
McCallie, J.M., 69, 70
McKelway, B., 153, 154
McPhee, J., 57
"Menacing moron" concept,
  42, 136
Mendel, Gregor, 55, 65, 173
Mental ability
  See also Intelligence
  and genetics, 21, 182
Mental retardation
  as genetic, 3, 4, 11, 74

voluntary sterilization of
  persons with, 174
*Mental Retardation: A Family
  Study*, 172
"Mercy" killings in Germany,
  164
Meyer, Adolph, 39, 40
Meyerson, Abraham, 66, 67,
  71, 76
Miller, Samuel, 1, 2
Mismeasure of Man, The, 79,
  83
Morons, 11, 12, 34, 41, 61, 116,
  136, 178
  detection of, 117
  incurability of, 34
  "menacing," 42, 136
Myers, W.S., 37
Myths, 8, 9, 16, 170, 193, 194

N

National Negro Tragedy, 181
Nature vs. nurture, 8
*Nature-Nurture Controversy,
  The*, 132
Nazis, 156, 157, 161, 169
  genocide by, 166
  race hygiene programs of, 7,
    156, 159
  sterilization programs of, 158,
    160
Nelson, K. Ray, 150, 151, 153
New eugenics, 8, 169–190
New York Times, 70
Nobel-laureate sperm bank,
  178, 179
Nuremberg Laws, 156
Nuremberg Trials, 8, 156
Nurture vs. nature, 8

O

*Official Handbook for
Schooling the Hitler Youth*,
165
Ohio State University, 64, 85
Olden, Marian S., 158, 159, 160

P

Pastore, Nicholas, 127, 132, 133
Pathological Institute of the
State of New York, 39
Patterson, Joseph Medill, 63, 64
Pedigrees in humans, 3
Pendell, Elmer, 179, 180
Pennsylvania State Normal
School, 39
Phalen, D., 45
Photo retouching in Kallikak
book, 79, 80, 83–84
Physically handicapped in
Germany, 164
Pine Barrens study, 57, 59
Plecker, W.A., 156, 157
Popenoe, Paul, 136, 137, 156
Popular appeal of Kallikak
book, 62
Popular Science Monthly, 65
Population's intellectual level
in America, 126
Porteus, Stanley, 67
Preface to Kallikak book, 49
Priddy, A.S., 148
Princeton University, 84, 126,
128, 129, 188
Prostitutes, 136
Prussian Ministry of Public
Health and Social Welfare,
159

*Psychology of the Normal and
Subnormal*, 128
Public opinion
about immigration, 127
about Kallikak book, 62

Q

Quaker Church (Society of
Friends), 39

R

Race, 138
and intelligence, 177
Race Betterment Foundation,
81
Race hygiene (*Rassenhygiene*)
of Nazis, 7, 156, 159
Racial differences, 172
Racial inferiority theories, 2,
180
Racial integration, 170, 172
Racially mixed families, 156
Randolph-Macon Women's
College, Department of
Psychology, 184
Rankin, Russell Bruce, 98
Rassenhygiene. See Race
hygiene
Recessive genes, 68, 73, 77
Reed, Elizabeth W., 172, 173,
174, 175
Reed, Sheldon, 172, 173, 174,
175
Reeves, Helen, 27, 29, 30, 31,
33, 34
*Release of the Destruction of
Life Devoid of Value, The*,
165

Reporting of neighbors under sterilization laws, 138–139

Reproduction control, 42, 43, 131, 136, 137

Research Institute of Human Development, Antioch College, 45

Research Institute, Temple University Medical School, 45

Research strategy of Goddard, 12

Restriction of child labor, 136

Restriction of immigration, 3, 7, 180

Restrictive marriage laws, 180

Retardation as genetic, 3, 4, 11, 74

Retouching of photos in Kallikak book, 79, 80, 83–84

Reviews of Kallikak book, 62, 65, 68, 75

Revolt Against Civilization, The, 3

Road to Survival, 175

Robbins, Emma Florence, 39

Robertson, G., 150

Rush, Benjamin, 84

Russians, 118, 120, 127

S

Sarason, Seymour, 189

Scarborough mansion, 37

Scheerenberger, R.C., 35

Scheinfeld, Amram, 67, 68, 70, 71, 72, 73, 75, 79

Schneider, Carl, 158, 167

Scholastic achievement, 182, 183

School desegregation, 172

Schulz, H., 35

Science, 71, 73

Scientific Monthly, The, 68, 69

Seed, The, 64, 65

Segregation of mentally inferior, 136, 137
   by colonization, 19
   by institutionalization, 19

Selden, Steven, 79, 80

Selective marriage, 2

Sex Versus Civilization, 179

Shelton, R.G., 148

Shirer, William, 164

Shockley, William, 176, 177, 178, 179, 180, 181, 182

Shuey, Audrey, 183, 184, 185, 186

Simon, Theodore, 41, 56

"Smith, Buck," 153, 154

Smith, Roy, 144, 145

Smithsonian Institution, 80

Snyder, L., 156

Snyderman, M., 127

Social Darwinism, 2, 8, 43

Social myths, 8, 9, 16, 170, 193, 194

Societal threat caused by hereditary defects, 42, 136

Society of Friends (Quaker Church), 39

Southern Europeans, 128, 138

"Special healing procedures" (*besondere Heilverfahren*), 164

Sperm bank for Nobel laureates, 178, 179

Standardized intelligence tests, 33, 117
   See also specific tests

Stanford University, 177

State Colony for Epileptics and Feeble-Minded, 139

Statue of Liberty, 115

Sterilization, 137, 139, 180, 182
  compulsory. *See*
  Compulsory sterilization
  "thinking exercise" about,
    177
  voluntary, 174, 176, 177, 178
Stockton family, 84
Stockton, Richard, 84
Stoddard, Lothrop, 3
Stroud, A.E., 142, 143, 144, 146,
  149
*Study of American
  Intelligence, The*, 127
Sunday Times of London, 188
"Super race" concept, 78
  See also Eugenics
Supreme Court ruling on
  sterilization laws, 139, 149,
  154

T

Teachers College, Columbia
  University, 132
Temple University Medical
  School Research Institute, 45
Terman, Lewis, 41, 73, 126
*Testing of Negro Intelligence,
  The*, 184
Tests
  See also specific tests, types
    of tests
  intelligence. *See* Intelligence
    tests
  standardized, 33, 117
  U.S. Army, 126, 127, 132
"Thinking exercise" about
  sterilization, 177
Thompson, Wayne, 167
Tragedy for American
  Negroes, The, 181

Training School for Feeble-
  Minded Girls and Boys,
  Vineland, New Jersey, 4, 6,
  11, 12, 15, 16, 21, 26, 27, 33,
  38, 39, 41, 42, 43, 56, 59, 64,
  69, 74, 79, 84, 85, 86, 98, 118,
  126, 129, 131, 160, 191
Twin studies, 186, 187, 188

U

Unionist-Gazette of Somerville,
  New Jersey, 100
Union Pacific Railroad, 42
Unit trait inheritance, 65–66, 75
University of Akron Archives of
  the History of American
  Psychology, 51, 73, 132, 160
University of California, 70
University of Heidelberg, 157,
  158
University of Minnesota, 172
University of Southern
  California, 39
University of Virginia, 151, 186
Unwed mothers, 136
U.S. Army tests, 126, 127, 132
U.S. Supreme Court ruling on
  sterilization laws, 139, 149,
  154

V

Van Wagenen, Blecker, 44, 63,
  64
Villanova University, 59
Vineland State School, 27, 35,
  86
Vineland Training School. *See*
  Training School for Feeble-

Minded Girls and Boys,
  Vineland, New Jersey
Virginia Act to Preserve Racial
  Integrity of 1924, 156, 157
Virginia Military Institute, 167
Vogt, William, 175, 176
Voluntary sterilization, 174,
  176, 177, 178
Voorhees, D.W., 38, 39, 45, 138
Voting, 130

W

Wallace, James H. Jr., 80
Wallin, J.E. Wallace, 27, 136
Walter Van Dyke Bingham
  Memorial Lecture, 187

Watson, Goodwin, 132
Wertham, F., 166
Western State Hospital,
  Staunton, Virginia, 140
Whitehead, 148
White, William Allen, 131
Wilhelm, Caroline, 147, 152
Wilker, Karl, 161, 162
Wolfensberger, W., 165, 166
Wood, Eula, 143, 144

Y

Yepsen, L.N., 70
Yerkes, Robert, 126
You and Heredity, 67, 76

About the Author

J. DAVID SMITH did his undergraduate work at Virginia Commonwealth University and received his doctorate from Columbia. A former Peace Corps volunteer, he has also taught special education and was presented with the Faculty Scholar of the Year Award at Lynchburg College. A member of the American Psychological Association and the American Association on Mental Deficiency, he has contributed to many professional journals in the field. *Minds Made Feeble* is his first book.